Advances in Anatomy
Embryology and Cell Biology

Vol. 84

Editors
F. Beck, Leicester W. Hild, Galveston
J. van Limborgh, Amsterdam R. Ortmann, Köln
J.E. Pauly, Little Rock T.H. Schiebler, Würzburg

F. Hajós   E. Bascó

The Surface-Contact Glia

With 25 Figures

Springer-Verlag
Berlin Heidelberg NewYork Tokyo
1984

Ferenc Hajós, M.D.
Eduardo Bascó, M.D.
1st Department of Anatomy
Semmelweis University, Medical School
Tüzoltó u. 58, Budapest H-1450
Hungary

ISBN-13:978-3-540-13243-1   e-ISBN-13:978-3-642-69623-7
DOI: 10.1007/978-3-642-69623-7

Library of Congress Cataloging in Publication Data
Hajós, F. (Ferenc), 1936–. Surface-contact glia.
(Advances in anatomy, embryology and cell biology; vol. 84) Bibliography: p.
1. Neuroglia. I. Bascó, E. (Eduardo), 1945–. II. Title. III. Series: Advances in
anatomy, embryology, and cell biology; v. 84. [DNLM: 1. Neuroglia – Physiology.
2. Cerebellum – Physiology. W1 AD433k v. 84/WL 102 H154s]
QL801.E67 vol. 84a [QL931] 574.4s 84-1247
ISBN-13:978-3-540-13243-1 (U.S.) [599'.0188]

© Springer-Verlag Berlin Heidelberg 1984

2121/3140-543210

Contents

1 General Introduction

1.1 Brief History

The diversity of cells constituting the central nervous system did not deceive last century neurohistologists in recognizing that this organ contained essentially two cell types: the nerve cells, or as termed according to the emerging concept of neural contiguity, the neurons, and the neuroglial cells. Neurons were clearly shown to be the means of excitability, impulse generation, impulse transmission, and connectivity in the neural tissue. The neuroglia, as indicated by its name ($\gamma\lambda\iota\alpha$ = cement or glue) given by Virchow (1860), was thought to be the cementing material ensuring the coherence of the nervous tissue, filling in the spaces of the neuropil, and isolating neuronal cell bodies. While this supposedly passive role did not attract multidisciplinary research on the neuroglia, successful efforts were made to extend our knowledge of the physiology, morphology, and biochemistry of neurons. As a result of this, the investigation of the neuroglia carried out in the first half of this century was mainly confined to morphology, often as a by-product of comprehensive analyses of neuronal systems. At any rate, the histological classification of the neuroglia was accomplished, laying a framework which has been used to the present day. Accordingly, the glia was divided into two major groups: the macro- and microglia. The former comprises two further subclasses, the astroglia and oligodendroglia. The microglia are still a matter of debate as far as its cell types and origin are concerned (Stensaas and Stensaas 1968; Vaughn and Peters 1968; Mori and Leblond 1969a; Peters et al. 1970; Privat and Leblond 1972). Since the late 1960s electron microscopic research has repeatedly suggested (King 1968) that the astroglia and oligodendroglia are two distinct groups of ectodermal origin, while the microglia comprises glial cells derived from the mesoderm. This view is at present neither generally accepted nor refuted.

A revival of interest in glial research began in the fifties. The introduction of powerful new techniques such as electron microscopy and, in particular, thymidine autoradiography has drawn attention to the fact that the neuroglia are more than a simple space-filling element of the nervous system. They were shown to be closely connected with important events of central and peripheral nervous system development [see Varon and Somjen (1979) for references]. Since then much has been done in ultrastructural histology and biochemistry, but to some extent also in physiology, to elucidate glial functions. The participation of the oligodendroglia in myelin formation was pointed out (Bunge 1968; Peters and Proskauer 1969; Peters and Vaughn 1970; Hirose and Bass 1973; Somjen and Trachtenberg 1979). On the basis of biochemical, physiological, and pharmacological evidence the astroglia are now believed to be involved in a complex metabolic interaction with neurons to maintain the ion balance of the extracellular brain space (Kuffler 1967; Somjen 1973, 1975; Katchalsky et al. 1974), to take up transmitters released into the extracellular space (Henn and Hamberger 1971; Hutchison et al. 1974; Snodgrass and Iversen 1974; Schrier and Thompson 1974; Schubert 1975), and to contribute to neuronal nutrition and energy metabolism (Svaetichin et al. 1965). With light and electron microscopy astroglial processes were demonstrated to be present at the sites of the blood-brain barrier, i.e., around brain capillaries and at some liquor-brain barrier sites such as pial surfaces (reviewed by Bunge 1970; Varon and Somjen 1979; Seress 1980). On this morphological basis, transport functions across various barrier systems were also attributed to this type of astroglia (Löfgren 1959; Kendall et al. 1969, 1972, 1973; Mitro 1974). Various types of astroglial processes are capable of phagocytosis [see Varon and Somjen (1979) for references], which is operational in removing degenerating neuronal elements. Massive proliferation of astroglial processes starts near brain injuries, called post-traumatic gliosis (Sjöstrand 1965, 1966a, b; Torvik and Skjörten 1971; Watson 1974; Davidoff 1973), and the glia form the scar after the healing of a brain wound (Cavanagh 1970). Kreutzberg (1966) was able to show astroglial proliferation during regeneration in the facial nucleus after the transection of the facial nerve. In another series of experiments it was pointed out that fine extensions of astroglial processes approach the synaptic junctions (Van der Loos 1963; Peters and Palay 1965; Špaček 1971; Hajós 1980), often completely isolating the synapse from its environment and making glia-neuron interaction feasible also at the synaptic level.

A number of clinical aspects have been raised by recent advances in glial research. Several neurological disorders have been shown to be caused by the disturbance of myelination, and the study of the oligodendroglia has therefore been more prominent, less attention having been given to the astroglia. Nevertheless, the latter type of glia also appears to be important in a number of clinical situations. The astroglia is the first affected by brain edema (Gerschenfeld et al. 1959); particularly the astroglia of the barrier sites react with severe swelling to agents and conditions causing brain edema (Klatzo et al. 1958; Raimondi et al. 1962). A substantial fraction of brain tumors is constituted by the gliomas, primarily astrocytomas. The tissue culture of glioma cells is a frequently used model to study pathologically increased cell replication and metabolism [see Wollemann (1974) and Giacobini et al. (1980) for references], but extrapolation of data obtained from these dedifferentiated cells to normal glial cells is questionable. Therefore, the understanding of normal glial development is gaining increasing importance with respect to tumor formation (Penfield 1931, 1932; Zimmerman 1955, 1957, 1969; Lynn et al. 1968; Lewis 1968b).

Further impetus to developmental studies of the glia has been given by the hypothesis of Rakic (1971a, b, 1972), who claimed that neuronal migration is guided by preexisting glial fibers. This would imply a primary role of the glia in the geometrical organization of regularly arranged brain areas such as the cerebral and cerebellar cortices, but in a broader sense the glia could be held responsible for a wide range of regularities in many brain areas that appear to develop their internal structural organization and neuronal wiring according to admirably determined programs. [For the most recent version of this hypothesis see Rakic (1981).] Indeed, the discovery of glial anomalies underlying genetic neurological disorders (Sotelo 1978; Caviness and Rakic 1978) supports the view that the normal development of the glia is essential in the normal development of the whole central nervous system.

1.2 Development of the Glia: Current Views and Problems

As it happened in the case of the adult brain, in the developing brain attention has also been primarily focused on neurogenesis rather than on gliogenesis. This is no surprise if we consider acceptable priorities. In spite of this, owing to the fact that the very roots of neurogenesis and gliogenesis are common, the early works of Kölliker (1896), Ramón y Cajal (1906, 1911), His (1904), and others, besides formulating the basic concepts of neurogenesis, also provided valuable information concerning the beginnings of gliogenesis.

What has clearly emerged from these pioneering studies is that the common source of neurons and neuroglia is the neuroepithelial wall of the neural tube and its derivative brain vesicles. This single layer of low cylindrical stem cells becomes stratified as a result of repeated mitoses and forms two additional layers called, according to current terminology (Boulder Committee 1970), the subventricular germinal zone and intermediate or mantle layer.

The first and still unresolved question arises at this point. Where does the earliest differentiation of the neuroepithelium into neuronal and glial cell lines occur? It seems that in the innermost layer cells are already differentiated into neuroblasts and so-called spongioblasts, which appear to be the glial precursors. Spongioblasts are asymmetrically bipolar cells having a central short process and a long one spanning the wall of the neural tube. Views on their further development are divergent. It is claimed that neurons are produced before their associated glial cells and the proliferation of glial cells occurs in relation to the differentiation and growth of neurons (Lorente de Nó 1933). As a result of advanced neurogenesis, spongioblasts lose their processes and either migrate outward to give rise to glioblasts and derivative glial cells or remain at their original site to form the ependyma, which is the lining of the central canal and brain vesicles (Soemmerring 1841; His 1887; Magini 1888a, b), and the subependymal glia.

There is, however, a population line of spongioblasts present during brain maturation that retain their original asymmetrically bipolar shape with long processes terminating in end feet at the outer surface (Golgi 1885; Lenhossék 1891; Retzius 1893, 1984; reviewed by Fleischauer 1972). Their derivatives were apparently not taken into consideration when formulating the rule of temporal relationship of neurogenesis and gliogenesis. The terminology of this cell reflects a great deal of theoretical confusion. The names ependymal tanycyte, ependymal spongioblast, ventricular cell, etc. and the more recent term of radial glia (Rakic 1971a, b, 1972) obviously portray the same cell type, which owing to its virtual disappearance from the mature central nervous system, has not been thoroughly investigated with respect to its possible derivative cells or cell lines.

Figure 1 shows three examples of the development of views on gliogenesis. The cell lineage proposed by Polak (1965) does not reckon at all with spongioblasts retaining their processes and thus serving as precursors of an additional cell line; Szentágothai (1977) mentions ependymal spongioblasts but suggests no relationship with the glial line; Varon and Somjen (1979) regard the late

No. 1. Polak (1965)

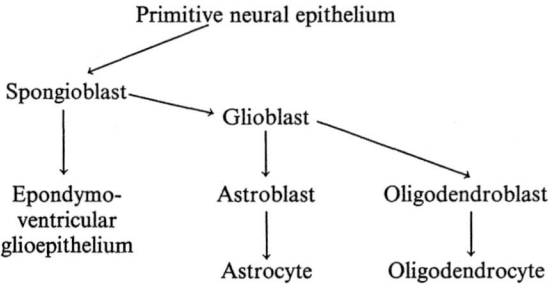

No. 2. Szentágothai (1977)

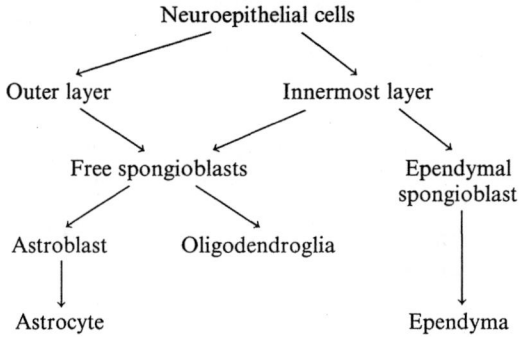

No. 3. Varon and Somjen (1979)

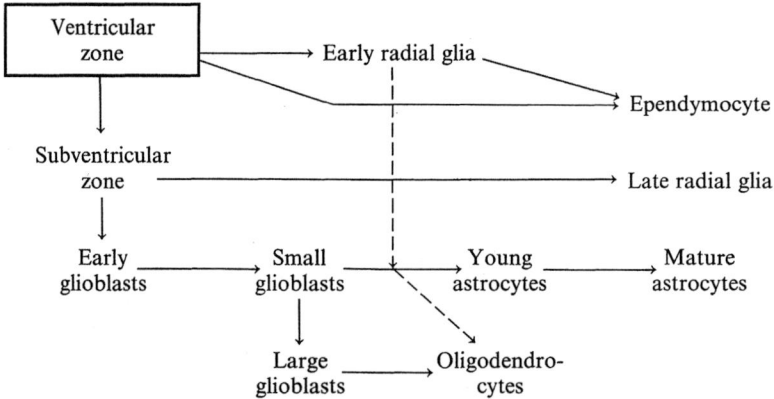

Fig. 1. Three current proposals for glial-cell lineages. *No. 1* does not reckon with spongioblasts retaining their processes and thus serving as precursors of an additional cell line. *No. 2* includes ependymal spongioblasts but suggests no relationship with the astrocyte line. *No. 3* comes closest to our views by regarding the late radial glia (i.e., persisting ependymal spongioblasts) as a distinct cell population, but no suggestion is made concerning its possible derivatives

radial glia (i.e., persisting ependymal spongioblasts) as a distinct cell population, but no suggestion is made concerning its further differentiation. These proposed cell lineages are typical examples of the process of recognition of persisting spongioblast-like cells during central nervous system maturation, but also of the inability to fit this population into any existing concept of gliogenesis.

Therefore, the basic problems dealt with in this study are connected with the cell line originating from these late spongioblast-like cells retaining processes that span the full width of the wall of the neural tube and brain vesicles. A number of data published in the past two decades (Fujita 1965a, b; Privat 1975; Sturrock 1976; Skoff et al. 1976a, b) suggest that the investigation of this embryonic glia may provide a clue to several poorly understood aspects of gliogenesis and related events of brain development such as:

1. The significance of postnatal mitotic activity at the nongerminal sites of the brain
2. The regional distribution and time course of glial proliferation in the brain
3. The temporal relationship of gliogenesis and neurogenesis
4. The participation of the glia in the structural organization of geometrically arranged brain areas
5. How to define the astroglia?

2 Materials and Methods

Animals. The pregnancies of B10/R and CBA strain mice were timed by testing vaginal smears. The 1st day after mating was counted as E1. Birth usually occurred on E20 or E21, and E21 was counted as the first postnatal day (P1).

Animals were kept on a standard laboratory diet with free access to water under 12-h alternating light and dark periods.

Resin Embedding for Light and Electron Microscopy. Animals from birth to 18 days of age were perfused under Nembutal anesthesia with Karnovsky's glutaraldehyde-paraformaldehyde fixative (Karnovsky 1964). The room-temperature fixative was perfused through a cannula inserted into the left ventricle of the heart. Brains were removed from the skull, halved in the midsagittal plane, and immersed overnight in Karnovsky's fixative at room temperature. Following a thorough washing in phosphate buffer (pH 7.4), halved brains were dehydrated *in toto* in graded ethanol series and propylene oxide, and embedded in Durcupan (Fluka). If the material was prepared for both light and electron microscopy the buffer washing was followed by a 2-h postfixation in 1% phosphate-buffered osmium tetroxide prior to dehydration and embedding. Frontal serial sections were cut from the whole hemisected brain at 10-μm intervals with a Tesla ultramicrotome and stained with 0.1%–1.0% toluidine-blue containing 1% sodium tetraborate to increase staining intensity. Durcupan was used as a mounting medium polymerized after coverslipping for 12 h in a 60° C oven.

Mitotic figures outside the germinal zones were mapped under the light microscope in each frontal section so as to provide an atlas of the distribution of sites of early postnatal proliferative activity not related to the germinal zones. To localize brain areas in the frontal sections, the anteroposterior distance from the bregma (corresponding to the frontal plane traversing through the anterior commissure) was calculated, taking into account section thickness, number of sections, and dimensions of the brain allowing for shrinkage in the embedding medium (Palkovits et al. 1971). It should be noted, however, that in contrast to paraffin embedding, where an approximately 25% shrinkage has to be reckoned with, shrinkage during resin embedding was less than 5%.

In the osmicated samples, territories were selected for ultrathin sectioning on the basis of the light microscopic examination of 1-μm-thick toluidine-blue-stained sections. Ultrathin sections of pale yellow and white interference colors were cut with Reichert and Tesla ultramicrotomes, floated on distilled water, picked up on uncoated copper grids, and stained with uranyl acetate and lead citrate (Reynolds 1963). Preparations were viewed and photographed under a Tesla BS 500 electron microscope.

6

Autoradiography for Light and Electron Microscopy. Animals were injected subcutaneously with a single dose of 10 µCi (=370 kBq/g body wt. [³H]thymidine [specific radioactivity 29 Ci (=1073 GBq)/mM] 1–12 days after birth. Animals were killed 1 h after the isotope injection either by decapitation or by the perfusion of Karnovsky's fixative through the left heart ventricle. Tissue blocks of the neocortex and hippocampus were excised, left immersed in Karnovsky's fixative overnight, washed in phosphate buffer (pH 7.4), and postfixed in 1% osmium tetroxide. Blocks were dehydrated through graded ethanol series and propylene oxide and embedded in Durcupan (Fluka).

For light microscopic autoradiography semithin sections were cut in the frontal plane, mounted on glass slides, and covered by dipping into Ilford K5 nuclear emulsion. After 42–46 days of exposure at +4° C, preparations were developed in Kodak D19 developer, fixed, and stained through the emulsion with 1% toluidine-blue containing 1% sodium tetraborate. Labeled cells with more than five grains were indicated as points on the projected drawing of each section examined.

Double-label autoradiography was performed in intact rats by applying [¹⁴C]thymidine (54 mCi/mM specific radioactivity) in addition to [³H]thymidine. The following experimental conditions were used.
1. A single dose of [³H]thymidine, 10 µCi/g body wt.
2. Five doses of 10 µCi/g body wt. [³H]thymidine administered at 6-h intervals
3. A single 10 µCi/g body wt. dose of [³H]thymidine followed 12, 20, and 24 h later by a single injection of 0.5 µCi/g [¹⁴C]thymidine
4. A single dose of 10 µCi/g [³H]thymidine followed 12, 20, and 24 h later by a series of five [¹⁴C]thymidine injections of 0.5 µCi/g, each at 6-h intervals starting 20 h after administration of the tritiated compound
5. A single 10 µCi/g body wt. dose of [³H]thymidine followed by a single injection of 5 µCi/g [¹⁴C]thymidine 2 or 4 h after administration of the tritiated compound

Animals were killed by decapitation 1 h after the last injection and brains were processed for autoradiography as described for the [³H]thymidine pulse label experiment.

For electron microscopic autoradiography the block was trimmed so as to contain the labeled cell selected under the light microscope. Ultrathin sections of yellow interference color were cut with Reichert and Tesla ultramicrotomes, floated on distilled water, and mounted on glass slides covered with a Parlodion film. Slides were dipped into Ilford L4 nuclear emulsion diluted to 1:4 with distilled water. Preparations were developed in Microdol-X after 20–40 days of exposure at +4° C, floated on distilled water, picked up on copper grids, and stained with uranyl acetate and lead citrate through the intact emulsion layer. Electron micrographs were taken with a Tesla BS 500 electron microscope.

Silver Impregnation. The rapid Golgi method (Palay and Chan-Palay 1974) was used en bloc. Brains were cut in the frontal plane into 100- to 300-µm sections by a Sorwall microchopper and covered with Durcupan (Fluka).

Immunocytochemistry of the Glial Fibrillary Acidic Protein (GFAP). The brains of postnatal pups (whole heads in the case of embryos) were snap frozen in isopentane in liquid nitrogen, and groups of four to six serial cryostat sections (12 μm thickness) were cut in the frontal plane at intervals of 250 μm through the entire cerebrum. After mounting onto formol-gelatin-coated glass slides they were dried on air and then fixed by 4% paraformaldehyde in phosphate-buffered saline (PBS; pH 7.4). This method proved to be the most satisfactory for preservation of both tissue structure and immunoreactivity. Following two PBS washes endogenous peroxidase activity was exhausted by incubation for 15 min in 0.3% H_2O_2. Sections were washed again and then incubated for 30 min in normal swine serum in 1:20 dilution, to block nonspecific staining. Preparations were then drained and incubated overnight in a 1:1000 dilution of rabbit anti-GFAP. The antiserum to GFAP, raised in the rabbit against GFAP prepared from the human spinal cord, was that described by Woodhams et al. (1980). Its specificity to the astroglia was tested on sections of adult rat cerebellum and it also showed a characteristic intracellular fibrillary staining pattern in cultured astrocytes. Swine anti-rabbit gamma-immunoglobulin (IgG) and rabbit peroxidase-antiperoxidase were both obtained from Dako (lot Nos. 069A and 079, respectively) and used at a dilution of 1:50. All antisera contained 0.25% Triton. Following three 10-min PBS washes the sections were incubated for 30 min in swine anti-rabbit IgG and after washing again in rabbit peroxidase-antiperoxidase. Sites of immunoreactivity were visualized by reaction for 5 min in 0.05% diaminobenzidine (DAB) in 0.05 M Tris buffer (pH 7.6) containing 50 μl/100 ml H_2O_2, the reaction being terminated by washing in distilled water. Sections were viewed either with hematoxylin counterstaining or without counterstaining.

Combined GFAP Immunocytochemistry and [³H]Thymidine Autoradiography. Pregnant mice (the timing of pregnancies was made as described on p. 6) were given a single intraperitoneal injection of 5 μCi (=185 kBq)/g body wt. [³H]thymidine (specific radioactivity, 10 Ci (=370 GBq)/mM at the same time of day, on E14, 15, 16, 17, 18, or 19. In one experimental series the animals were killed 15 h after the isotope injection, while in another they were left until birth (E21) before the brains were processed for GFAP immunostaining as described on p. 7. Following extensive washing in distilled water to remove any remaining diaminobenzidine, the slides were dried and stored at room temperature before dipping. Autoradiography was carried out by dipping into Ilford K5 nuclear emulsion diluted 1:2 with distilled water. After an exposure time of 14–25 days at +4° C they were developed in Kodak D19L developer, fixed, and lightly counterstained with hematoxylin to reveal cell nuclei before mounting in DPX.

Horseradish Peroxidase Tracer Method. The skulls of newborn, 4-, 7-, 12-day-old, and adult mice were opened under Nembutal anesthesia. The dura and arachnoid were carefully opened so as to expose the pia. A glass capillary was then introduced near the exposed pial surface by means of a stereotaxic apparatus. The glass capillary was connected to a microsyringe by a plastic tube. A drop of 30% solution of horse radish peroxidase (HRP) (Sigma VI) in physiological saline was deposited on the brain surface with the utmost care to avoid damaging the pia. Following the deposition of HRP the skull was left open for 10 min. Animals were killed after a 48-h survival period by trans-aortic perfusion, first of a physiological saline solution to remove blood, then of an aldehyde fixative. For the light microscopic visualization of HRP, 20-μm frozen sections were cut from the cerebrum and cerebellum in the frontal and sagittal planes, respectively, and processed as described by Mesulam (1976). Sections were counterstained with safranin and photographed using either bright- or dark-field condensers.

3 Postnatal Cell Proliferation at Nongerminal Sites of the Brain

3.1 Introductory Remarks

Postnatal cell-proliferation in the central nervous system has been well established since the beginning of this century (Allen 1912; Smart 1961; Altman 1962, 1966a, b; Altman and Das 1967; Angevine 1965; Taber Pierce 1966, 1967, 1973). In mammals most of the neurons are formed during embryonic life; only a few of them continue proliferating after birth, particularly in the forebrain. In the mouse forebrain the granule cells of the olfactory bulb and hippocampal dentate gyrus are generated from the last third of gestation until day 20 after birth (Altman and Das 1965; Altman 1966b; Angevine 1965; Hinds 1968a, b). During the course of normal development postnatal proliferation of other forebrain neurons has not been verified. Even for these areas the early postnatal period is the latest time of neuronal proliferation; this is followed by their differentiation, after which, as stated by Greenfield et al. (1958), they undergo no further divisions.

The rodent cerebellum is somewhat exceptional in this respect because its neuronal proliferation continues until P28–30 [see Jacobson (1978) for references].

In contrast to the neurons the neuroglia appear to retain much of their proliferative capacity throughout life. Early this century Allen (1912) suggested the death and formation of glial cells in the adult brain. This has been confirmed over the past decades by autoradiography, demonstrating postnatal gliogenesis in the brains of rat, mouse (Smart and Leblond 1961; Altman 1962, 1966a, b; Hinds 1968a, b; Dalton et al. 1973; Sturrock 1974a, b), cat (Fleischauer 1966, 1968; Hang 1972), and rabbit (Robain 1970).

As mentioned in Sect. 1.2, the common germinal sites for both the neurons and neuroglia are the subventricular zones. To these join some so-called secondary germinal sites such as the external granular layer of the immature cerebellar cortex, the granular layer of the hippocampal dentate gyrus, and the internal part of the olfactory bulb granular layer, where prolonged neurogenesis takes place over considerable postnatal periods. The retained proliferative capacity of glial cells is manifested in mitotic activity outside the primary and secondary germinal layers.

3.2 Mitotic Activity at Nongerminal Sites of the Immature Cerebellar Cortex

In the immature cerebellum, we observed mitotic figures in 1-µm sagittal sections of the vermis in the upper part of the granular layer of the cerebellar cortex near the layer of Purkinje cells between P7–12 (Fig. 2). They were much larger than the mitotic figures of the external granular layer, which gives rise to granule neurons. Mitotic cells showed clear-cut connections with the radial fiber system of the primitive molecular layer. This fiber system continued through the external granular layer up to the pial surface. Fibers had the typical "watery" appearance of the glia, i.e., a very light appearance, as if the structure had been "washed out." They seemed to originate from the apical pole of the mitotic cells, only a few processes being directed toward the depth of the internal granular layer. The pattern of their arborization observed in the sagittal plane was similar to that characteristic for the Golgi picture of Bergmann glia cells (Palay and Chan-Palay 1974; Das 1976; Hajós et al. 1982). In most cases mitoses took place in a plane parallel to the surface as judged by the orientation of mitotic spindles. Mitotic cells were thus identified as the Bergmann glia.

3.3 Mitotic Activity at Nongerminal Sites of the Immature Forebrain: Time Course and Regional Distribution

We mapped mitotic figures at nongerminal sites of the forebrain in the early postnatal period using 1-µm frontal serial sections of the whole forebrain stained with toluidine blue. The laboriousness of the technique as compared with conventional paraffin histology was compensated for by a better visibility of mitotic figures in all developmental stages studied.

The fact that our mitotic cells belonged to a subpopulation contained by developing brain areas caused major difficulties in expressing their numbers in reliable quantitative terms. The calculation of mitotic index is possible either in a homogeneous cell population or in relation to a stable cell type used as a reference. Neither of these references are present in the developing forebrain except for the corpus callosum and other commissural bundles, where only glial cells occur. This situation has been exploited for the quantitative study of glial proliferation in these areas (Sturrock 1974a, b, 1976). In the developing forebrain regions the cell population is a mixture of undifferentiated neurons and glial cells, which have not yet been identified. A further problem is that cell packing-density is different in the regions studied and is continuously altered by the growth of the brain. Hence the expression of mitotic figures per unit area is of no comparative value.

In this situation it would have been unrealistic to stress any known means of quantitation. We performed semiquantitative estimation instead, with the restriction that differences were evaluated only in the course of development of the same area. Accordingly, to denote the degree of mitotic activity, plus signs (+) were used and the term "massive" used for the highest intensity of mitotic activity (Table 1).

Numerous mitotic figures were demonstrated at nongerminal sites of the forebrain from birth until day 12. They were located in the neuropil or in close vicinity to large neurons. The cytoplasm of mitotic cells was large, irregularly shaped with a typical light appearance. Mitotic cells in the vascular system and in the meninges were clearly distinguishable by their smaller size and darker cytoplasm, and were not evaluated.

At birth (P1, Fig. 3) medium mitotic activity was found in the frontal cortex, low in the pyriform cortex. The hippocampal formation and the corpus callosum showed massive proliferation.

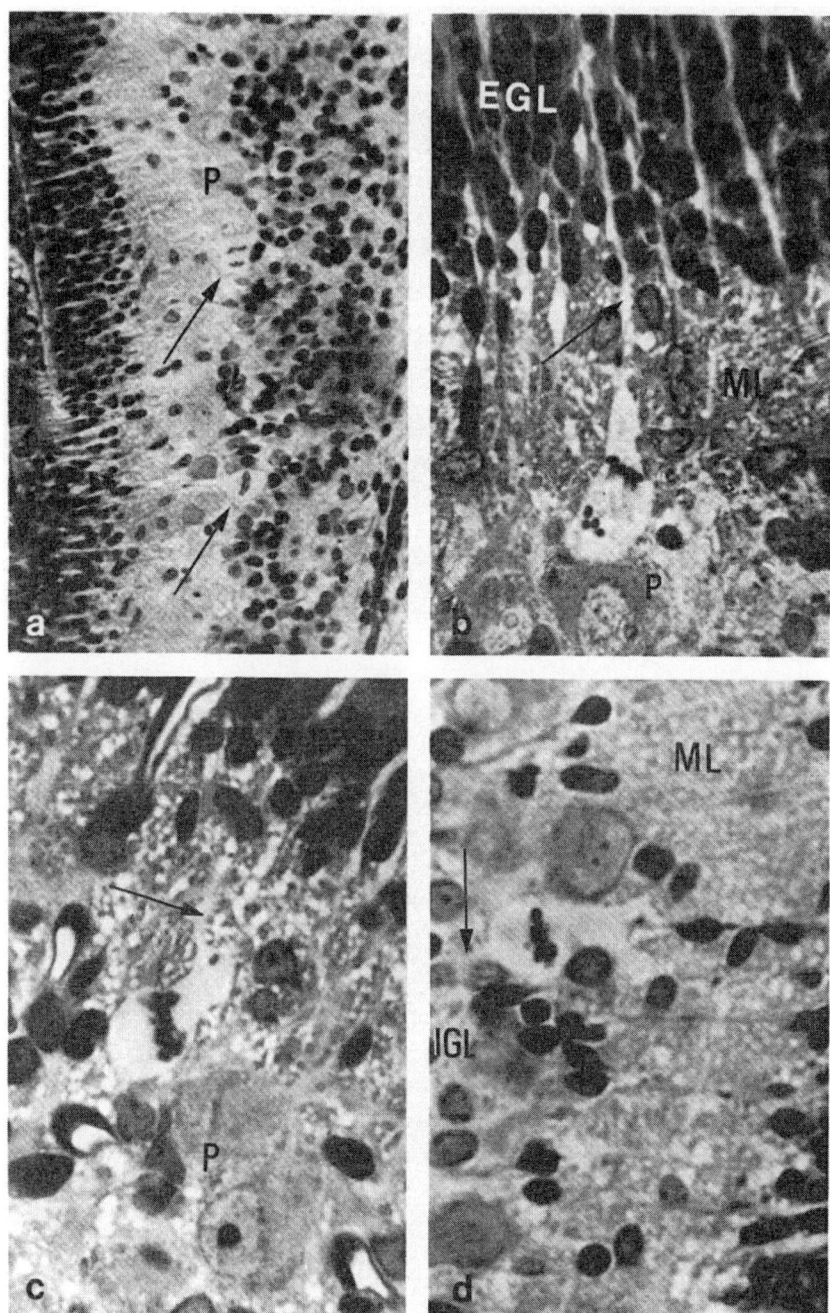

Fig. 2a–d. Mitotic cerebellar Bergmann glia cells in the 10-day-old rat. **a** Survey micrograph of the cerebellar cortex with mitotic cells (*arrows*) in the ganglionic layer. One of the mitotic cells has a process directed toward the pial surface. Note the orientation of the mitotic spindle: in the lower cell it is parallel, while in the upper one it is perpendicular to the surface. *P*, Purkinje cells. × 250. **b** Mitotic cell with a radial fiber (*arrow*) extending to the pial surface through the primitive molecular (*ML*) and external granular (*EGL*) layers. *P*, immature Purkinje cells. × 1200. **c** Mitotic cell with radial processes (*arrow*) showing a ramification pattern

Table 1. Regional distribution and intensity of mitotic activity in the immature mouse forebrain

Brain area	Age in days					
	0	3	5	7	10	12
Frontal cortex	+ + +	+ + +	Massive	+ + +	+	−
Sensorimotor cortex	−	+	Massive	Massive	−	−
Occipital cortex	−	−	+	+ +	−	−
Pyriform cortex	+ +	+ +	+	+	+	+
Hippocampus and dentate gyrus	Massive	Massive	Massive	Massive	+ +	+ +
Corpus callosum	Massive	Massive	−	−	−	−
Limbic areas	−	+ + +	+	+	+	+
Caudate putamen	−	Massive	−	+	−	−

Crosses indicate estimated intensity of mitotic activity

On P3 (Fig. 4) massive proliferation continued in the hippocampal formation and corpus callosum, accompanied by medium activity in the frontal cortex and in the septal nuclei and amygdala. The latter two are henceforth referred to as the limbic areas. The sensorimotor cortex contained sporadic mitotic figures. The caudate-putamen, which showed no mitotic activity at birth, was found to proliferate massively.

On P5 (Fig. 5) in the frontal and sensorimotor cortices numerous cell divisions were found, while in the occipital and pyriform cortices mitotic activity was very low. The number of mitotic figures was invariably high in the hippocampal formation and very low in the limbic areas. The sudden fall in proliferative activity of the caudate-putamen and corpus callosum at this stage, as compared with the previous stage, was remarkable.

On P7 (Fig. 6) a decrease occurred in the number of mitotic figures in the frontal cortex, whereas massive cell multiplication continued in the sensorimotor cortex. Divisions in the occipital cortex showed an increasing number but still within the low range. Mitotic activity in the pyriform cortex was very low. The hippocampal formation was still proliferating massively. The limbic areas and caudate-putamen contained sporadic mitotic figures.

On P10 (Fig. 7) the mitotic activity of the frontal cortex decreased to a very low level, while that of the sensorimotor and occipital cortices ceased. Very low mitotic activity continued in the pyriform cortex and limbic areas. A significant decrease from massive to low was observed in the proliferative activity of hippocampal formation.

On P12 (Fig. 8) the distribution and number of mitoses were virtually unchanged as compared with P10. By this age early postnatal mitotic activity appeared to come to a standstill.

During the postnatal period studied no cell divisions were seen in the diencephalon.

---

typical for Bergmann glia. The mitotic spindle is parallel to the surface. *P*, Purkinje cells. × 1200. **d** Mitotic cell with the origination of radial processes at its pole toward the molecular layer (*ML*) and of a thin process (*arrow*) directed toward the internal granular layer (IGL). The mitotic spindle is parallel to the surface. × 1200

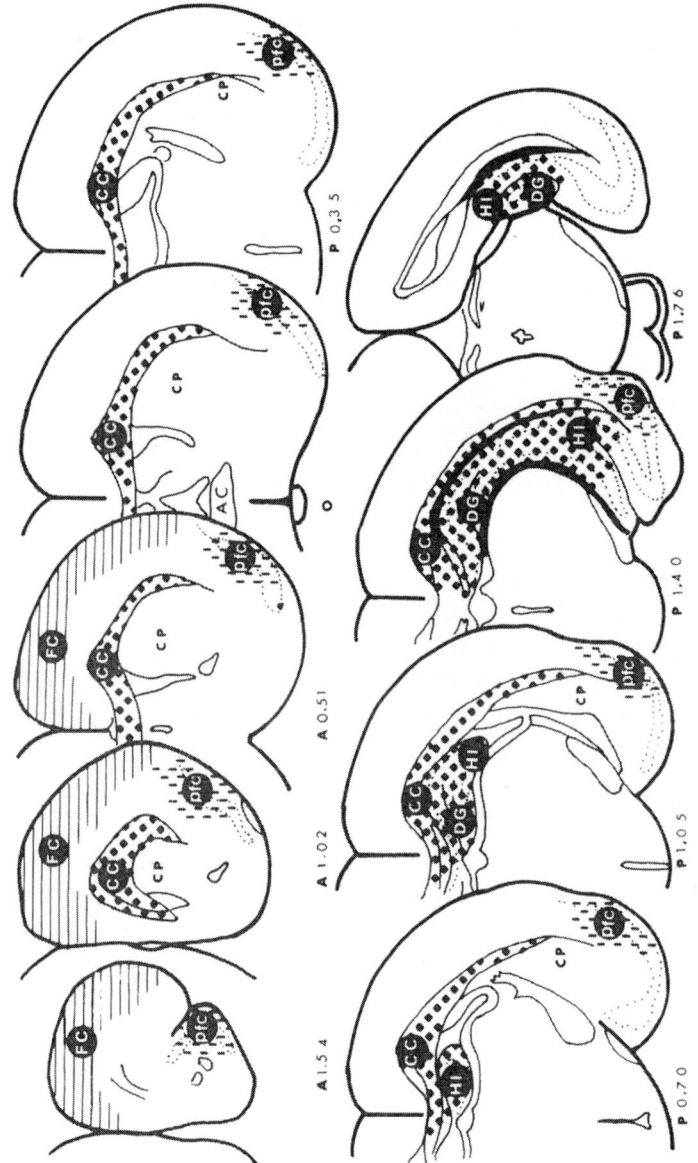

Fig. 3. Newborn mouse (P1)

Figs. 3–8. An atlas of the regional distribution and intensity of mitotic activity in the immature mouse forebrain. Frontal serial sections of epoxy resin-embedded and toluidine-blue-stained forebrains are represented by schematic drawings. Indication of intensity of mitotic activity: ▨, massive; ▩, medium; ▨, low. Values *A* and *P* represent distances in millimeters anterior and posterior, respectively, to the bregma (the frontal plane through the anterior commissure). *AC*, anterior commissure; *CC*, corpus callosum; *CP*, caudate putamen; *DG*, dentate gyrus; *FC*, frontal cortex; *HI*, hippocampus; *LA*, limbic areas; *pfc*, pyriform cortex; *SM*, sensorimotor cortex

14

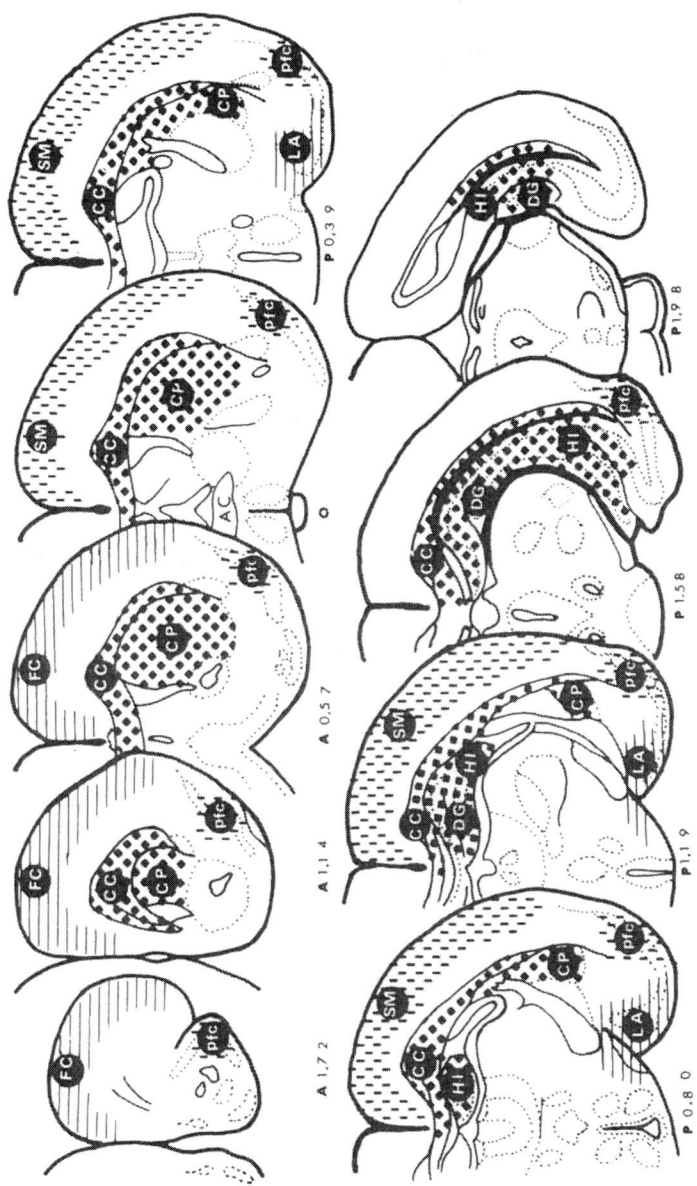

Fig. 4. Three-day-old mouse (P3)

3.4 Comments

3.4.1 Cerebellum

Proliferating cells have repeatedly been observed in the internal granular layer of the developing cerebellum (Ramón y Cajal 1929; Miale and Sidman 1961; Fujita et al. 1966; Lewis et al. 1977). With the aid of routine histological tech-

15

Fig. 5. Five-day-old mouse (P5)

niques mitotic figures have been seen until the 12–14th postnatal days (Lewis et al. 1977). During this period intensive labeling of numerous internal granular layer cells with [³H]thymidine has also been demonstrated (Miale and Sidman 1961; Fujita et al. 1966). It has been concluded that this proliferative activity corresponded to the multiplication of the cerebellar astroglia (Miale and Sidman 1961; Lewis et al. 1977). It is less clear what the relationship is between proliferating internal granule cells and the abundant glial fiber system of the primitive molecular layer. These fibers are characteristically perpendicular to the cerebel-

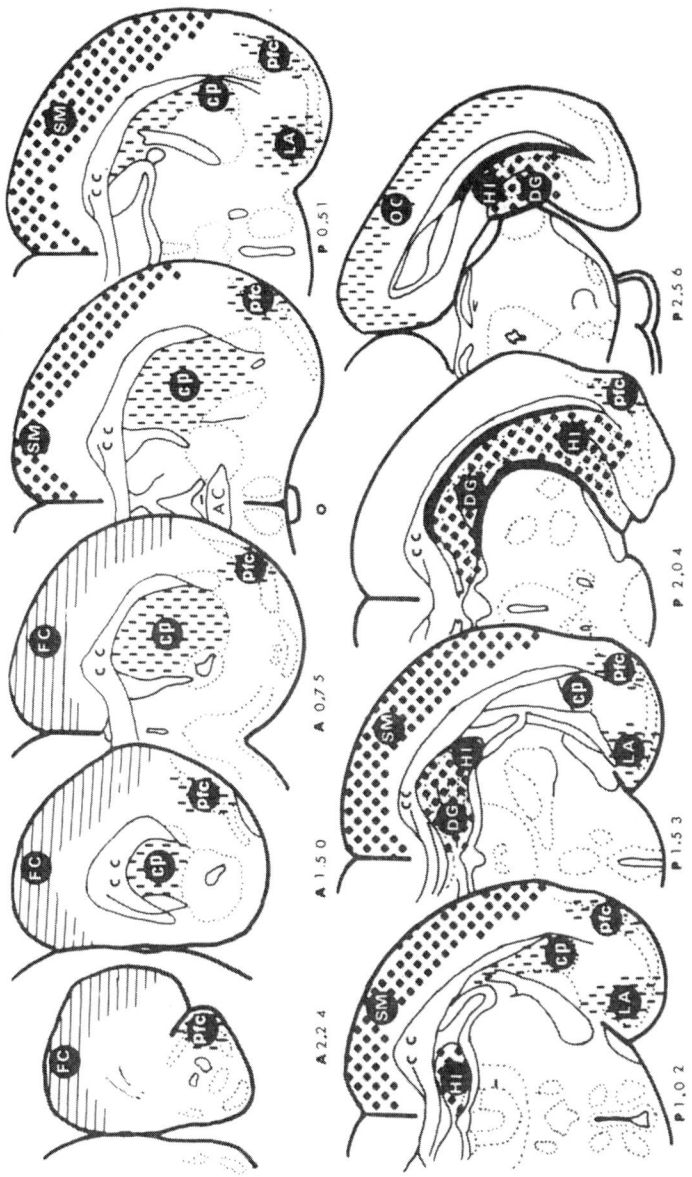

Fig. 6. Seven-day-old mouse (P7)

lar surface. They were observed soon after birth by Ramón y Cajal (1890). It is also of interest that from P7 the mitoses as well as [^{3}H]thymidine-incorporating cells of the internal granular layer (IGL) were observed to become gradually restricted to the layer of Purkinje cells (Miale and Sidman 1961; Lewis et al. 1977). On this basis the glial nature of these mitotic cells was suggested (Lewis et al. 1977).

There is a great deal of controversy concerning the generation time and development of the Bergmann glia. In the mature cerebellar cortex this special type

17

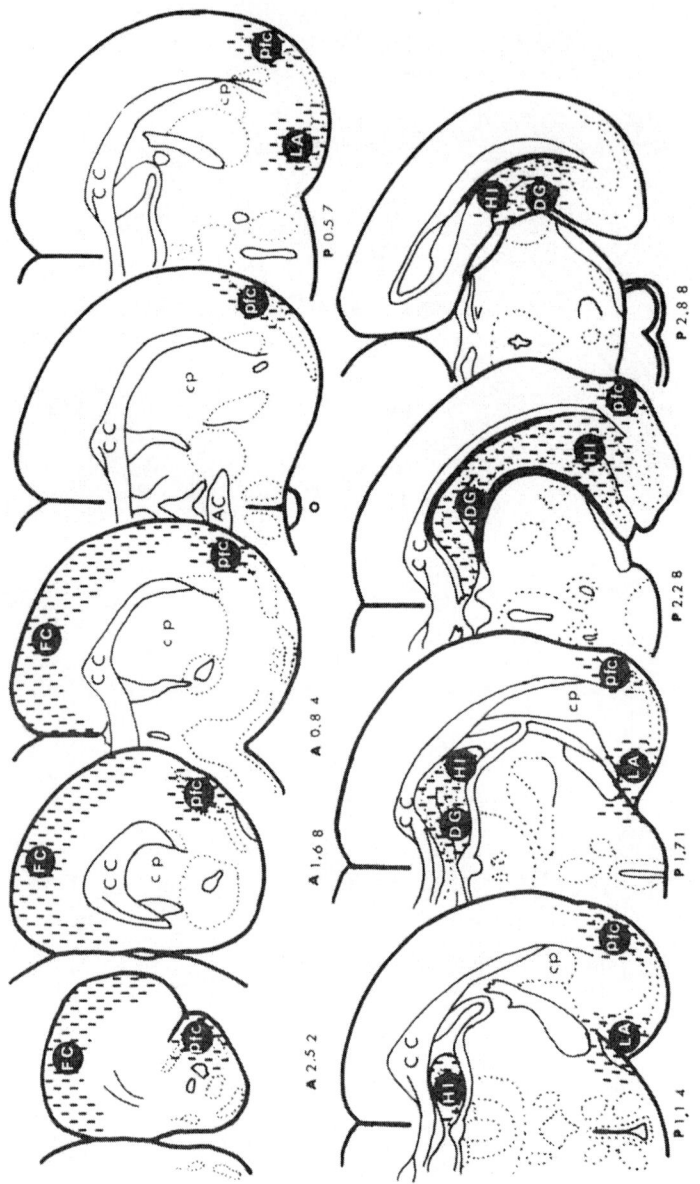

Fig. 7. Ten-day-old mouse (P10)

of astroglia is characterized by processes traversing the molecular layer perpendicularly to the cerebellar surface and the cell bodies located in the ganglionic layer. Because of its apparent participation in the organization of the neatly geometrical structure of the molecular layer the development of the Bergmann glia has been paid special attention (Fujita et al. 1966; Bignami and Dahl 1973, 1974b; Das et al. 1974; Altman 1975; Das 1976). It seems, however, that the divergent views on generation time are partly due to different interpretations of basically unconflicting results. With silver impregnation techniques glial cells

18

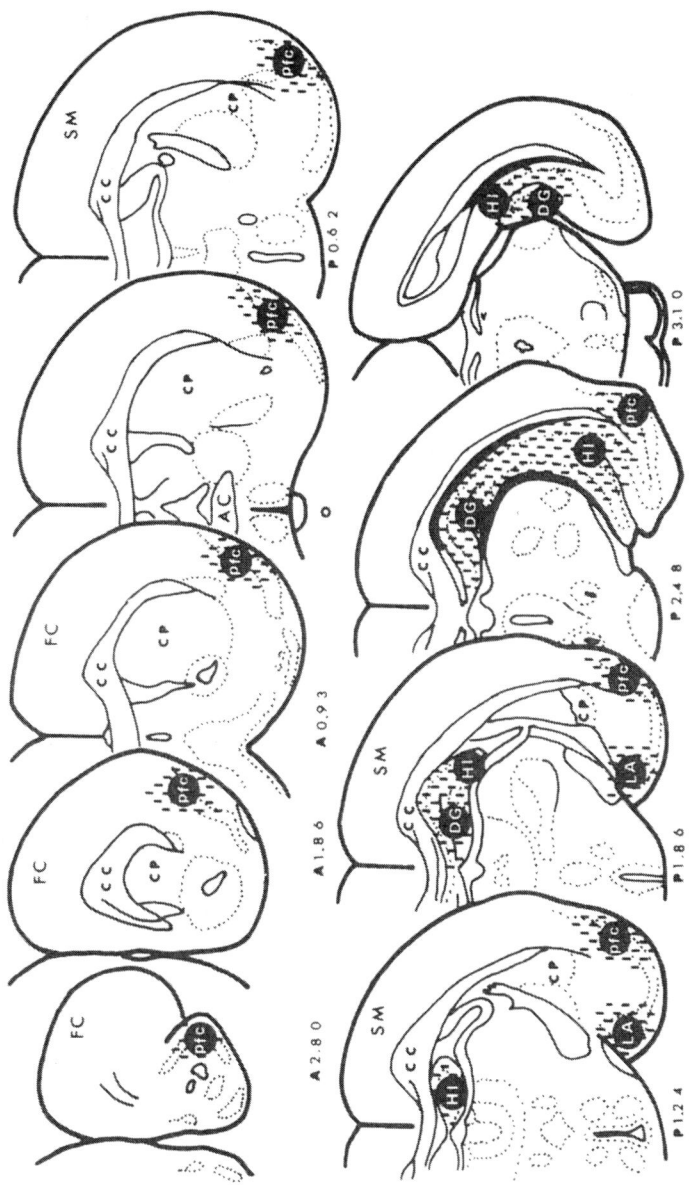

Fig. 8. Twelve-day-old mouse (P12)

having the radial fibers could be shown from the first postnatal days (Ramón y Cajal 1890). With electron microscopy these fibers could be verified (Rakic 1971 b) in the primitive molecular layer. However, the intensive incorporation of [³H]-thymidine into cells lying near to the Purkinje cells between 9 and 12 days led Das et al. (1974) to conclude that the peak rate of formation of Bergmann glia falls into this period. Our findings suggest that at this time there is indeed increased mitotic activity around the Purkinje cells, but that mitoses occur in cells with their typical radial fibers already developed (Fig. 2).

This unusual phenomenon means that radial glial fibers might appear well before the proliferative activity of their cell bodies has ceased. This leads us to suppose that the cell bodies where the radial fibers of the first postnatal week originated were among the proliferating cells in the middle zone of the IGL. These cells possessing radial fibers would eventually occupy a position near the Purkinje cells. If we assume that Bergmann glia cells are more essentially characterized by their radial fibers than by the position of their cell bodies, an explanation can be offered to resolve discrepancies concerning generation time. It appears that the proliferative activity observed between 9 and 12 days around the Purkinje cells does not principally differ from that observed at earlier stages in deeper parts of the IGL: both are in fact mitoses occurring in glial cells, some of which had previously extended their processes to the cerebellar surface. Between 9 and 12 days the last series of mitoses takes place, involving glial cells which by this time have occupied the typical position of mature Bergmann glia.

The orientation of the mitotic spindles indicates that developing Bergmann glia cells might split either perpendicular to the surface, thus sharing the radial fibers or, what seems to be more common, in a plane parallel to the surface. In the latter case one of the daughter cells may inherit the whole radial fiber system, whereas the processes of the other daughter cell would be directed toward the depth of the IGL. These cells could develop new radial fibers or serve as periglomerular astrocytes in the mature granular layer.

3.4.2 Forebrain

In the areas studied, except for the dentate gyrus, no postnatal neurogenesis has been verified. Therefore, the mitotic figures we have observed from birth up to P12 were attributed to glial cell multiplication. The oligodendroglia known to be involved in myelin formation is generated around the end of the second postnatal week [see Jacobson (1978) for references]; hence mitotic figures observed before this age are most likely to be astroglial precursors. This is supported also by the large size, elongated shape, and light cytoplasmic staining of mitotic cells unlike the appearance of mitotic figures of both neuroblasts and the oligodendroglia. These properties are thought to characterize histologically mitotic astroglial elements (Cavanagh 1970). However, mitosis is only a stage of the cell cycle when cells undergo marked morphological alterations. Their observation in a nonmitotic period and the further confirmation of their astroglial nature we attempted by [³H]thymidine autoradiography.

3.5 Light and Electron Microscopic Description of [³H]Thymidine-Labeled Cells at Nongerminal Sites of the Postnatal Brain

3.5.1 Light Microscopic Autoradiography

Cerebellum. With pulse-labeling with [³H]thymidine from birth to P12, we were able to reveal intensive proliferative activity from birth up to days 10–12 in the internal granular layer. Around P6 (Fig. 9) labeled cells were distributed

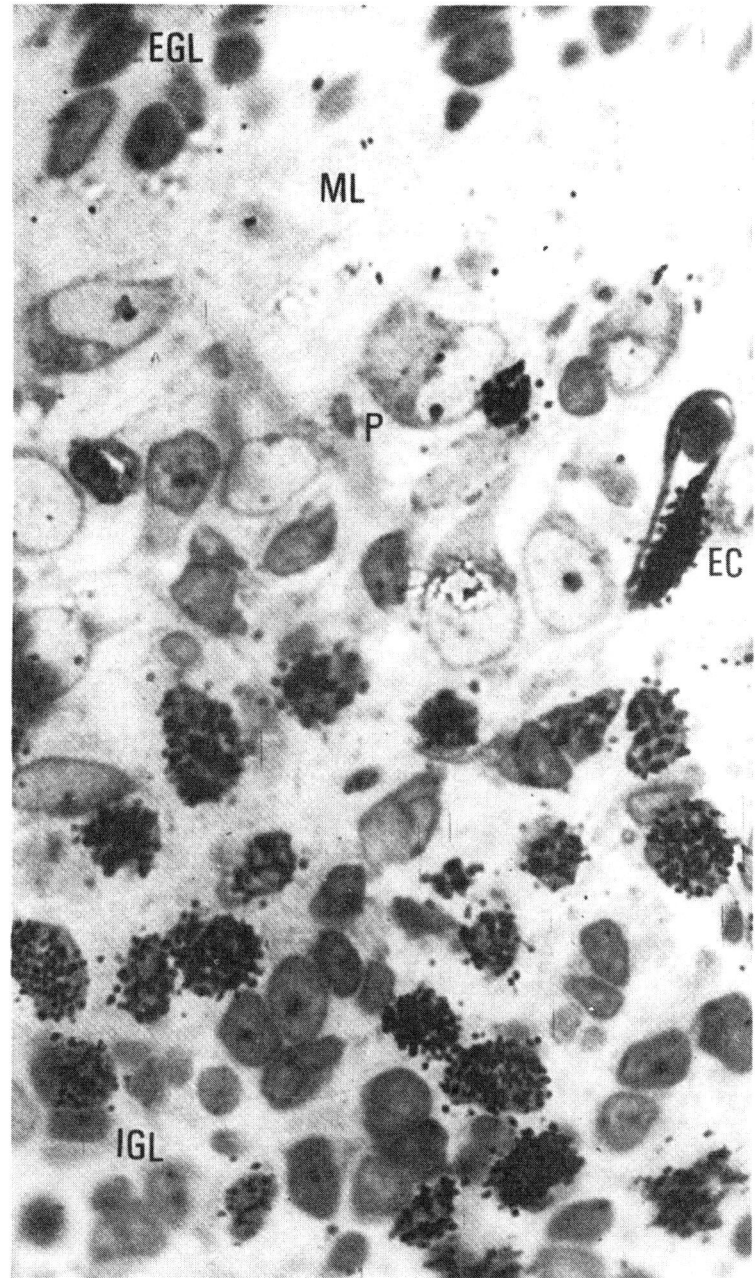

Fig. 9. Pulse-labeling with [³H]thymidine of the cerebellar cortex at P6. Purkinje cells (*P*) are still irregularly arranged. In the prospective internal granular layer (*IGL*) pale-nucleated cells are seen, many of them covered by silver grains due to radioactivity. Labeled cells are distributed throughout the full width of the internal granular layer. *EC*, proliferating endothelial cell; *ML*, primitive molecular layer; *EGL*, external granular layer. × 1700

throughout the full width of the developing internal granular layer. With increasing age they became gradually restricted to the upper zone of the internal granular layer, while from days 9–10 (Fig. 10) label was observed exclusively over cells situated in the uppermost cell-row of the internal granular layer or in the ganglionic layer. The assessment of morphology of labeled cells was generally hampered by overlying silver grains. However, in the more advanced stages (from P7) the number of silver grains was reduced due to the dilution of label by repeated mitoses, and in these cases labeled cells proved to be identical to the pale nucleated cells of the internal granular layer identified by Lewis et al. (1977) as glial precursors.

After P12 no [^{3}H]thymidine incorporation was seen either in the internal granular or in the ganglionic layers. In the double-isotope labeling experiments ^{3}H, ^{14}C, and double-labeled cells could be distinguished by the different isotope uptake periods. Conforming to general experience, cells accumulating [^{3}H]thymidine were characterized by densely packed silver grains over their nuclei, those incorporating [^{14}C]thymidine showed a scattered, widespread granulation, whereas the double-labeled cells had densely granulated nuclei surrounded by a loose halo of silver grains. The difference in grain distribution after ^{3}H and ^{14}C was apparent at equal concentrations of the isotopes. This difference was enhanced by applying [^{3}H]thymidine at a 80-fold higher concentration than the ^{14}C-labeled compound.

Following a single injection of [^{3}H]thymidine, 12% of internal granular layer cells were found labeled. If the whole proliferative pool was labeled by repeated injections of the isotope a 25% value was obtained. The peak rate of entry into a new S-phase occurred 20–24 h after the injection of [^{3}H]thymidine, as revealed by successive [^{14}C]-thymidine injections at various intervals. This type of isotope treatment also showed that the cells did not enter the next DNA-synthesizing cycle synchronously. The 20- to 24-h peak interval seemed, however, to be a suitable starting time to show the total proliferative pool of cells repeatedly synthesizing DNA. Accordingly, a 12% fraction of the labeled population proved to renew synthesis of DNA after 20–24 h. The length of the S-phase was determined on the basis of the above findings using the equation:

$$S = h \frac{(^3H + {}^{14}C) + {}^3H}{{}^3H}$$

proposed by Korr et al. (1973). Calculations were performed for the ^{14}C injections at 2 and 4 h; both calculations yielded approximately 6 h for the duration of the S-phase.

Forebrain. Using [^{3}H]thymidine autoradiography, proliferating cells were pulse-labeled at P1, P3, P7, P10, and P12. From P3, labeled cells were found scattered in a distribution roughly similar to that showed for mitotic figures (Sect. 3.2). A more detailed analysis was carried out in the cerebral cortex, where under oil immersion the characteristic nuclear structure of labeled cells could be revealed.

In one type of labeled cell the nucleus was banana shaped with evenly distributed, pale chromatin (Fig. 11 a). Nuclei were in an eccentric position near the nuclear envelope, which had a thin rim of accumulated chromatin. The nucleus

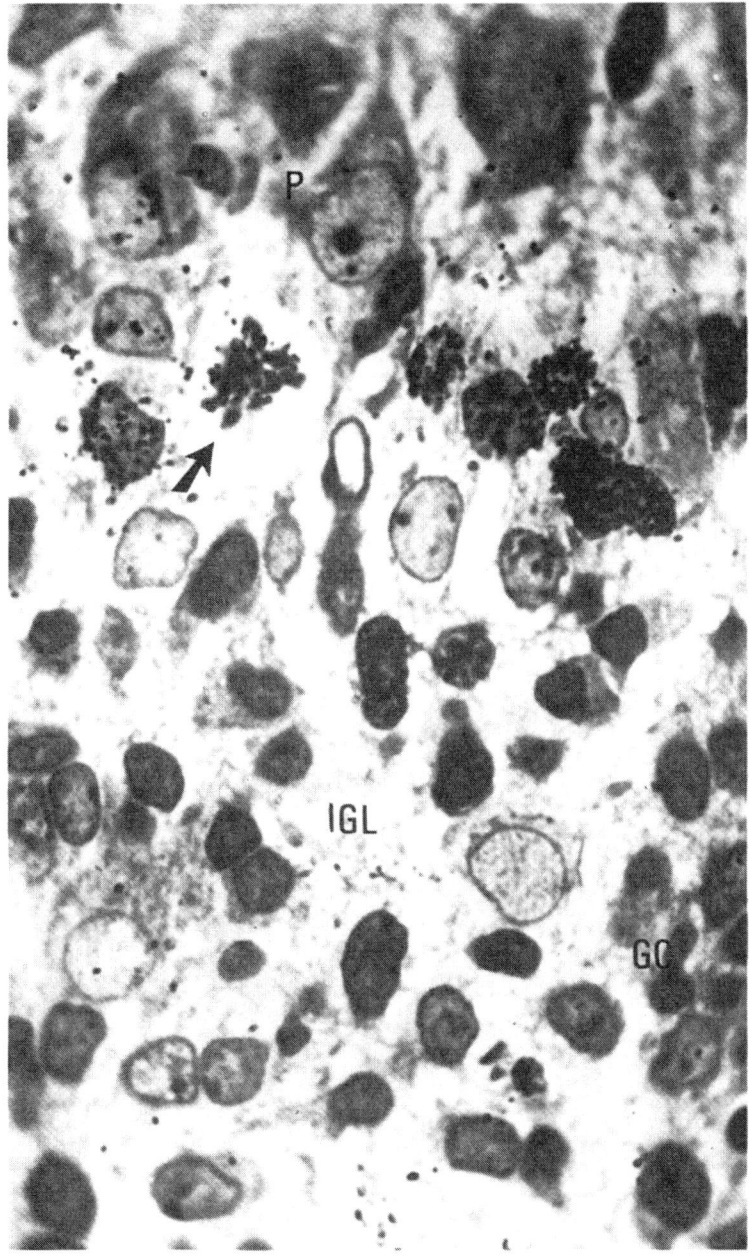

Fig. 10. Pulse-labeling with [³H]thymidine of the cerebellar cortex at P10. Label is observed exclusively in the upper part of the internal granular layer (*IGL*) near the Purkinje cells (*P*). A mitotic cell (*arrow*) is also labeled. Pale-nucleated cells of the internal granular layer are reduced in number while granular cells (*GC*) are frequently seen. × 1700

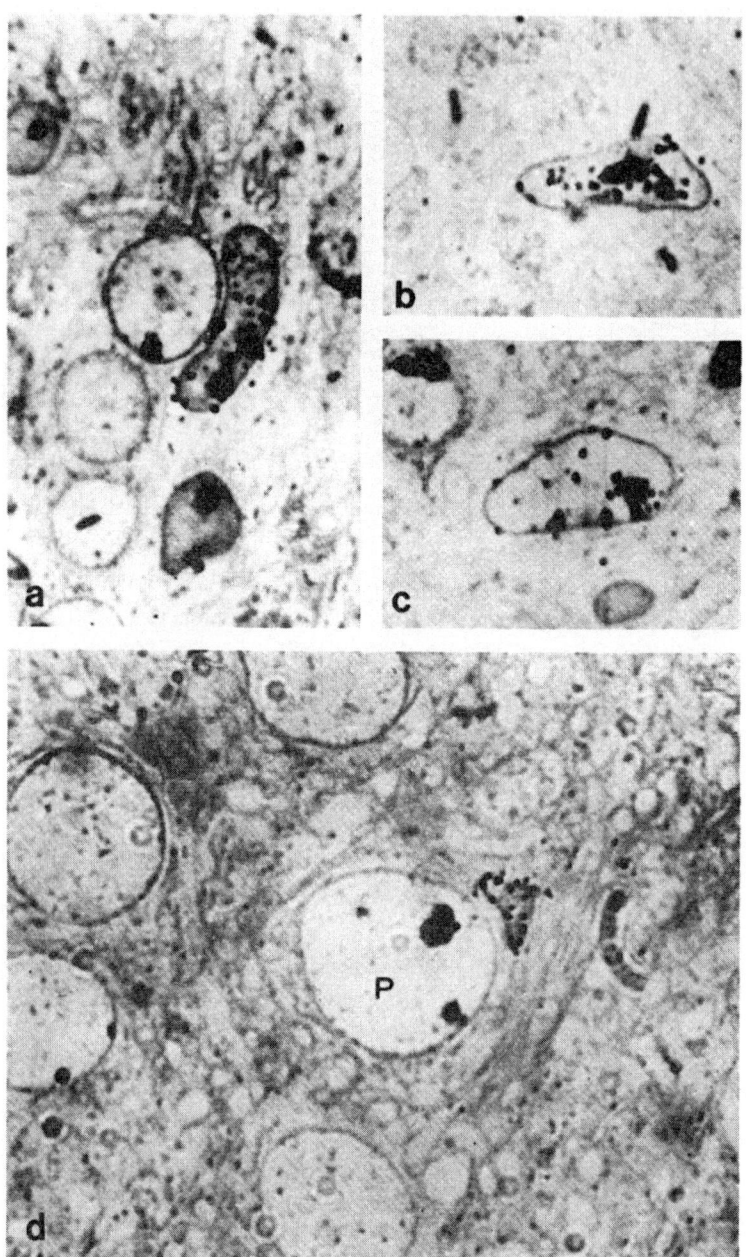

Fig. 11a–d. [^{3}H]Thymidine-labeled cells in the developing cerebral cortex of the 7-day-old mouse. **a** Labeled cell with pale, banana-shaped nucleus, prominent nucleolus, and evenly distributed chromatin forming a thin apposition at the nuclear envelope. × 1700. **b, c** Labeled cell with light nucleus and prominent nucleolus. The shape of the nucleus resembles that of a footprint. × 1700. **d** Labeled cell in satellite position to a pyramidal cell (*P*). × 1700

of the second labeled cell type resembled a footprint (Fig. 11 b, c), with an internal structure similar to that of the first type, with the exception that the nucleoli also occurred in a central position. Both the "banana" and "footprint" nuclei were aligned vertically to the pial surface, suggesting that they were in the migratory state. A third type of labeled cell was found in a satellite position to large neurons (Fig. 11 d). It was smaller than the former two two types and the nucleolus was less conspicuous. All labeled cell types had a characteristically light cytoplasm from which extensions could be followed for some distance in the surrounding neuropil.

3.5.2 Electron Microscopic Autoradiography

The glial nature of cells proliferating in the cerebellar internal granular layer from birth to P10–12 has been well established with light microscopic technique (Lewis et al. 1977; Bascó et al. 1977). In the forebrain the differentiation between early glioblasts and neuroblasts is rather difficult (Caley and Maxwell 1968; Meller et al. 1968; Sturrock 1976, 1978; Wolff and Rickmann 1977; Imamoto et al. 1978; Raedler and Raedler 1978; Robain and Ponsot 1978). It has been shown that glial precursors undergo repeated mitoses prior to differentiation (Paterson et al. 1973; Korr et al. 1975; Moskovkin et al. 1978) and that the last divisions occur near the final position of the cell (Hommes and Leblond 1969; Das et al. 1974; Skoff et al. 1976a; Bascó et al. 1977; Mares and Brückner 1978; Moskovkin et al. 1978).

We carried out electron microscopic autoradiography in the developing cerebral cortex to identify glial precursors the proliferation of which was light microscopically verified. We studied their ultrastructure in the early postnatal (P3–12) period. The parietal cortex and the hilus of the dentate gyrus were chosen for study because these two areas differ principally in the course of neurogenesis. In the neocortex, neurogenesis terminates before birth (Berry 1974; Brückner et al. 1976; Lund and Mustari 1977; Reznikov et al. 1978), while in the dentate gyrus more than 80% of the granule neurons are acquired postnatally (Bayer and Altman 1974; Schlessinger et al. 1975; Stanfield and Cowan 1979). We thought therefore that the investigation of these areas would yield information also concerning the relationship of neurogenesis and gliogenesis.

Under the electron microscope two types of labeled cells were observed in the cerebral cortex. The first (Fig. 12a) was small, occasionally ovoid. The nucleus contained clumps of chromatin subjacent to the nuclear envelope. The cytoplasm was poor in organelles; mainly scattered free ribosomes occurred. The other labeled cell type (Fig. 12b) was larger with a round or polygonal nucleus. Nucleoli were often prominent and chromatin was sparse, evenly distributed and forming a thin apposition at the inner aspect of the nuclear envelope. The cytoplasm appeared to be immature as judged by the low number of organelles and scattered free ribosomes.

In the dentate gyrus of 3- to 5-day-old mice, labeled cells were seen under the electron microscope throughout the whole gyrus, whereas by days 10–12 they were confined to the plexiform layer. Some of them formed a neatly continuous row under the granular layer (Fig. 13) in a typical position of astrocytes of the mature dentate gyrus (Woodhams et al. 1981). Labeled cells in this subgranular position were of two types, similar to those in the parietal cortex.

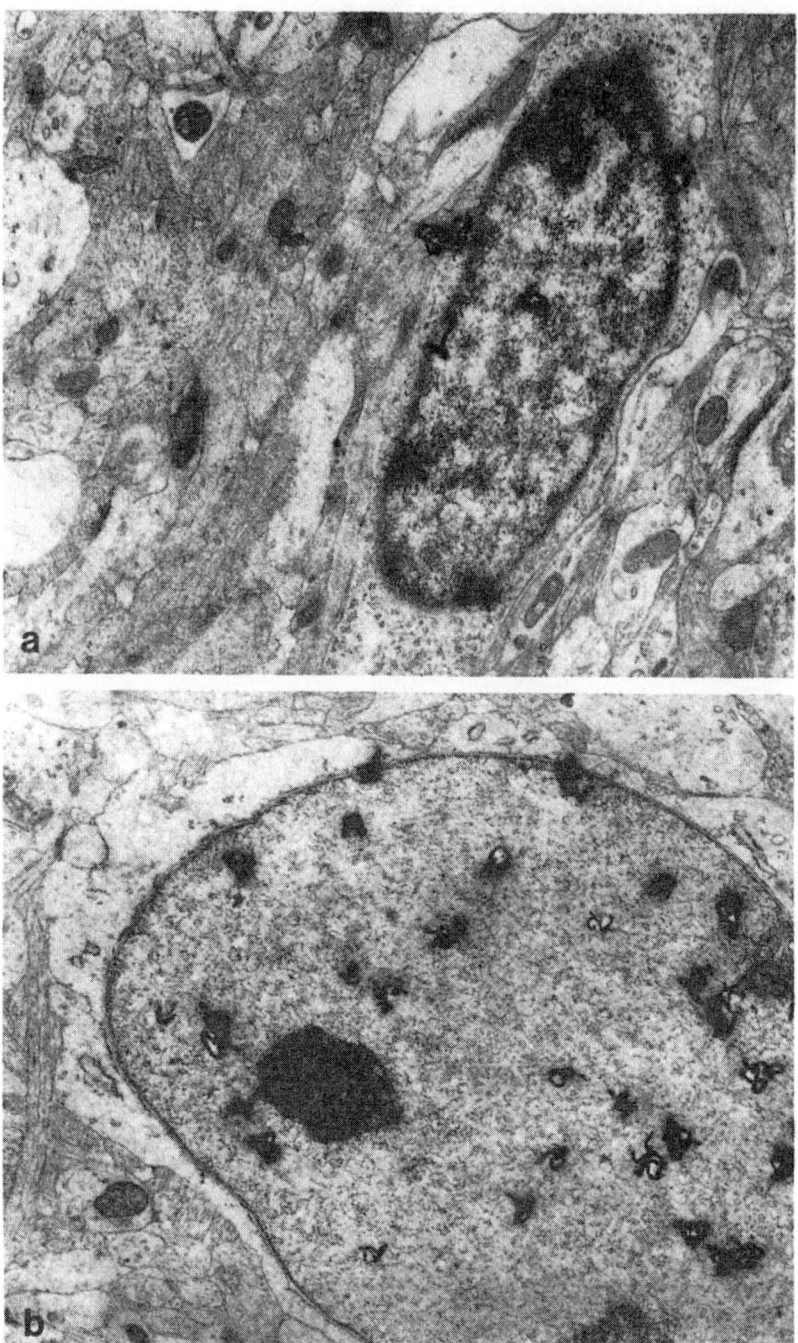

Fig. 12a, b. Electron micrographs of [³H]thymidine-labeled cells in the cerebral cortex of the 5-day-old mouse. **a** Small labeled cell. The nucleus contains clumps of chromatin pronouncedly subjacent to the nuclear envelope. The cytoplasm is poor in organelles; mainly scattered free ribosomes are seen. × 12000. **b** Large labeled cell. The nucleus is nearly round with sparse, evenly distributed chromatin forming a thin apposition at the inner aspect of the nuclear envelope. The cytoplasm is of low electron density and particularly poor in organelles. A few scattered free ribosomes and endoplasmic sacs are visible. × 12000

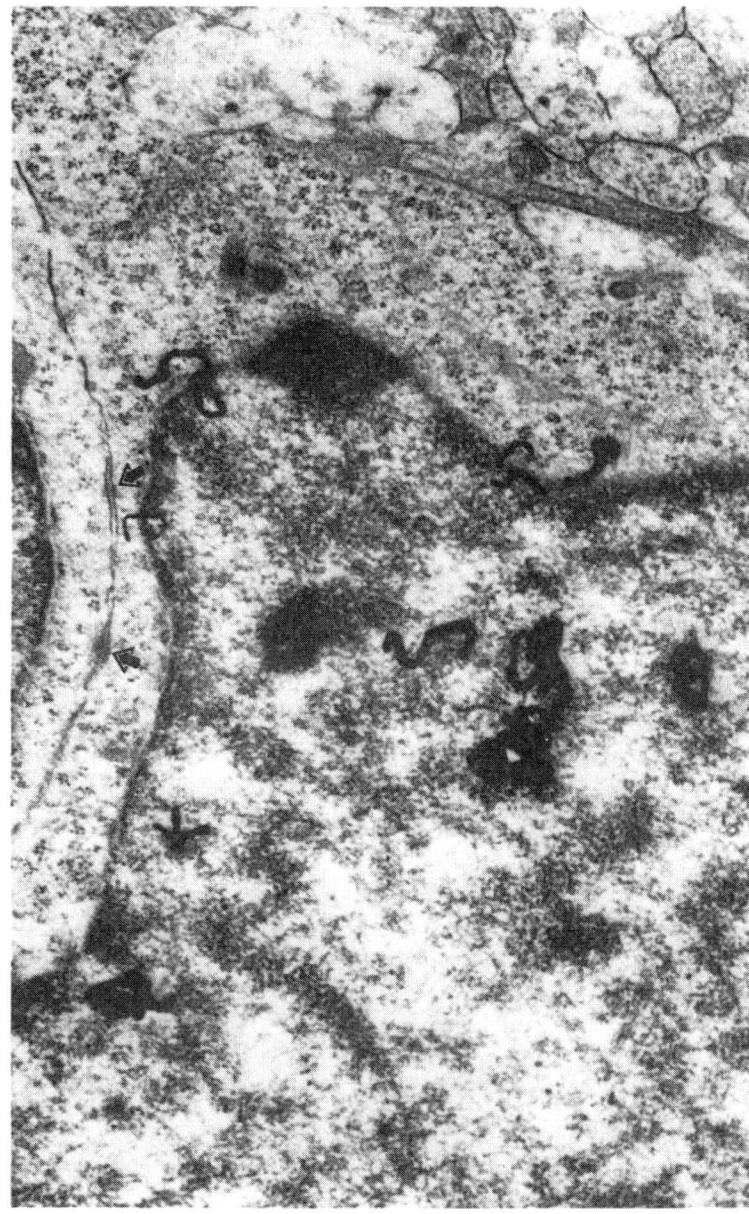

Fig. 13. Electron micrograph of [³H]thymidine-labeled cells from the subgranular layer of the 12-day-old mouse dentate gyrus. Between neighboring labeled cells desmosome-like junctions are present (*arrows*). The cytoplasm of labeled cells is of low electron density and is poor in organelles. × 38 000

In both areas studied nuclei of the large cells (Fig. 12b) resembled astrocyte nuclei (except for the prominent nucleolus) and so did the cytoplasm by containing only few organelles and having an electron-lucent appearance.

Either in the neocortex or in the dentate gyrus, labeled cells, if contacting each other, had numerous desmosomoid junctions between their adjoining surfaces (Fig. 13).

3.6 Comments

3.6.1 Cerebellum

Data presented and the findings of Lewis et al. (1977) and Fülöp et al. (1979) indicate that the cell population of the internal granular layer of the early postnatal cerebellar cortex consists mainly of proliferating Bergmann glia precursors.

This situation offered a model suitable for determining some parameters of proliferation of a population of glial cells having radial processes.

As indicated by the parameters calculated from double-label experiments, the proliferation intensity of the immature glial population of the internal granular layer was comparable to that of neurons (Lewis and Lai 1974; Lewis et al. 1977). As to the 6-h length of the S-phase obtained by the double-label technique, it has to be noted that although this value corresponds to those obtained by other workers using this technique in different organs (Maurer et al. 1972; Schultze et al. 1973) it appears to be shorter than the S-phase calculated on the basis of labeled mitoses. The discrepancy between the two methods is well known (Maurer et al. 1972; Schultze et al. 1973); its discussion is beyond the scope of the present work.

The injection time of [^{14}C]thymidine was chosen to rule out repeated labelings of the same cell within one S-phase. The time elapsing between the two isotope injections (20–24 h) was shorter than the migration time of external granular layer cells (Fujita 1967). Therefore, such cells, even if they took up the double label, could not reach the internal granular layer. The double labeling has clearly shown that the isotope-incorporating cells synthesize DNA inside the internal granular layer.

3.6.2 Forebrain

Extensive studies of glial development have been carried out only in the white matter of the forebrain (Mori and Leblond 1969a, b, 1970; Privat and Leblond 1972; Ling and Leblond 1973; Paterson et al. 1973; Sturrock 1974a, b, c, 1976; Imamoto et al. 1978) and in the optic nerve (Vaughn and Peters 1967; Vaughn 1969; Skoff et al. 1976a, b). The electron microscopic autoradiographic studies of Skoff et al. (1976a) and of Imamoto et al. (1978) have identified three types of proliferating glial cells. Sturrock (1974a, b, c), in his description of glial elements in the developing anterior commissure, distinguished four types of immature glial cells, three of them identified as precursor cells. Less is known of glial development in the gray matter of the central nervous system. Autoradiographic studies have verified postnatal proliferation of the glia in the cerebral cortex and hippocampus (Smart and Leblond 1961; Reznikov 1968; Hommes and Leblond 1969; Altman 1968; Mares and Lodin 1974; Bayer and Altman 1974; Mares and Brückner 1978; Reznikov et al. 1978), but owing to the limitations of light microscopic autoradiography the safe identification of early developmental stages was not possible. The present electron microscopic autoradiographic study added to earlier light microscopic and routine electron microscopic observations by verifying also in the cerebral cortex and hippocampus the existence and proliferative nature of at least two of the three types of glial precursors suggested by Sturrock (1976). On the basis of ultrastructural landmarks these cells of the cortex and dentate gyrus correspond to the small and large glioblasts found in the corpus callosum (Sturrock 1976). It is difficult to decide whether these types belong to the same cell line or represent, as claimed by Sturrock (1976), respective maturational stages of the astrocyte and oligodendrocyte lines. The cytoplasmic and nuclear ultrastructure and the presence of desmosomoid contacts would indicate that according to the criteria of Skoff et al. (1976a) large glioblasts may be proliferating forms of the astrocyte line in a fairly advanced stage of differentiation. On the other hand, small proliferating glioblasts appeared to be undifferentiated. Their differentiation into glial precursors could be predicted on the basis of their localization, number, and timing of proliferation rather than specific ultrastructure. The fact that the peak formation of the oligodendroglia in the parietal cortex takes place later (cf. Jacobson 1978) argues for the assumption that both our large and small labeled cells are early precursors of astrocytes. Comparing the light and electron microscopic findings it seems that the labeled cells seen between P3 and P12 under the light microscope at various nongerminal sites correspond to the large type of glioblasts, presumably astroblasts.

When comparing the cortex and the dentate gyrus no fundamental difference was found in the morphology of the two types of glioblasts. While in the cortex postnatal neurogenesis is negligible, the granular layer of the dentate gyrus retains its neurogenetic capacity over a long postnatal period (Bayer and Altman 1974; Schlessinger et al. 1975; Stanfield and Cowan 1979). The similarity of the pattern of glial proliferation in the two regions suggests that postnatal glial proliferation, at least at these sites, is independent of neurogenesis.

In the following we shall consider the relationship between postnatal, nongerminal proliferative activity — attributed, on the basis of what has been experimentally shown and discussed, to gliogenesis — and the radial glia.

4 Radial Glia in the Pre- and Postnatal Brain

4.1 Introductory Remarks

The morphological similarity between the cerebellar Bergmann glia and the embryonic radial glia suggests a possible developmental relationship, even in the adult cerebellum. Less obvious is such a relationship in the mature forebrain where embryonic radial fibers virtually do not persist. Therefore, in the following the forebrain radial fiber system and its fate during development shall be our concern rather than its counterpart in the cerebellum, where quite a number of studies have been carried out on this issue (Rakic 1971b; Bignani and Dahl 1973).

Ventricular cells whose processes at early embryonic stages span the full width of the brain move to the ventricular zone at a later stage, while still contacting the outer surface with a long process. They retain their connections with the ventricular lumen by a short process. This asymmetrically bipolar cell type was described last century by the silver-impregnation studies of the brain (see Sect. 1!). For its general designation the term "radial glia" has recently been proposed by Rakic (1971a, b, 1972). There remained, however, ambiguities concerning the true glial nature of this cell type as, for example, bipolar neurons are also encountered during this embryonic period.

Recent advances in immunocytochemistry of glial cells (Bignani and Dahl 1973, 1974a, b, 1975) have enabled a more comprehensive investigation of the radial glia to be performed. Evidence has been presented for the prenatal presence of typical radial fibers (Del Cerro and Swarz 1976; Swarz and Oster-Granite 1978; Choi and Lapham 1978). Less is known about the regional distribution, postnatal persistence, and mechanism of disappearance of the radial glia.

We have investigated these questions by means of the immunocytochemical demonstration of GFAP in the developing forebrain with comparative silver-impregnation studies.

4.2 GFAP Immunocytochemistry of the Developing Forebrain

GFAP immunostaining was looked at in the mouse brain from E14 until adult-hood.

4.2.1 Cerebral Cortex

Radially oriented immunopositive fibers were seen first in the frontal cingulate cortex at E17. Radial fibers spanned the telencephalon, projecting from the upper-medial wall of the lateral ventricle to the interhemispheric fissure and formed an immunopositive plexus near the pial surface. At E18 (Fig. 14a–d) immunoreactive radial fibers spanning the full width of the telencephalic wall were found in the upper frontal cortex. At higher magnifications (Fig. 14c, d) they could be traced from the ventricular to the pial surface. In agreement with previous reports (e.g., Levitt and Rakic 1980), inside the telencephalon a close association between glial fibers and migrating young neurons was observed (Fig. 14b). Rows of migrating cells attached to radial fibers occurred in the telencephalic white matter whereas in the cortical plate the dense packing of cell bodies obscured this type of association between migrating cells and the radial glia.

The bulk of neurogenesis and nerve cell migration occurs prenatally in the mouse cortex (Angevine and Sidman 1962), but the appearance of immunopositive radial fibers continued after birth. Both this and their subsequent gradual disappearance (which followed a pattern similar to their development) were observed to be regular in time course and regional distribution. While by about P4 well-developed radial fibers were evident both in the cingulate and more laterally in the upper parietal cortices (Fig. 15a–d), at P7 they could no longer be detected in the anterior cingulate cortex. Only short subpial processes were seen here throughout the remainder of the period studied, and they were particularly pronounced at the depth of the interhemispheric fissure. At P7 most radial fibers were found to originate from the subventricular zone of the ventrolateral angle of the lateral ventricle adjacent to the stria terminalis (Fig. 15), and from this site they projected to the middle and ventral parietal regions. This projection persisted over a considerable postnatal period (Fig. 15c).

By P14 (Fig. 15b, d) immunopositive radial fibers were seen in the ventral half of the parietal cortex, being absent in other parts of the neocortex apart from the area around the stria terminalis. This rostrocaudal and dorsoventral gradient of appearance and disappearance of immunopositive fibers continued so that after P20 the area around the stria terminalis was the only place where the origin and initial emanation of GFAP-positive, radially directed fibers was consistently detected. In anterior (frontal) cortical areas the dorsoventral gradient of development was not so marked and radial fibers appeared earlier than in more posterior regions: at P4 they were present throughout the frontal cortex.

During the whole developmental period studied radial fibers appeared to have a slightly undulating straight course and a brush-like terminal ramification under the pia (Fig. 15d), although because of the folding of the extreme margins of the sections the external glia limitans often appeared artifactually thickened.

A consistent finding in the present study was the lack of correlation between the pattern of formation of cortical plate and the appearance of GFAP-positive radial fibers. Separation of the cortical plate was already evident at E14, and by E15 it was present as a distinct layer in the lateral aspect of the cerebral vesicles, extending dorsally to a level just above the primordium of the striatum. Study of serial coronal sections revealed no obvious differences between anterior and posterior levels in the time of appearance of the cortical plate. By E16 the cortical plate was well developed up to the lateral border of the cingulate cortex (i.e., the interhemispheric fissure), and by E17 it extended round the cingulate cortex to the depth of the interhemispheric fissure (Fig. 15a, b). At this age the only region of the cortex where the cortical plate could not be

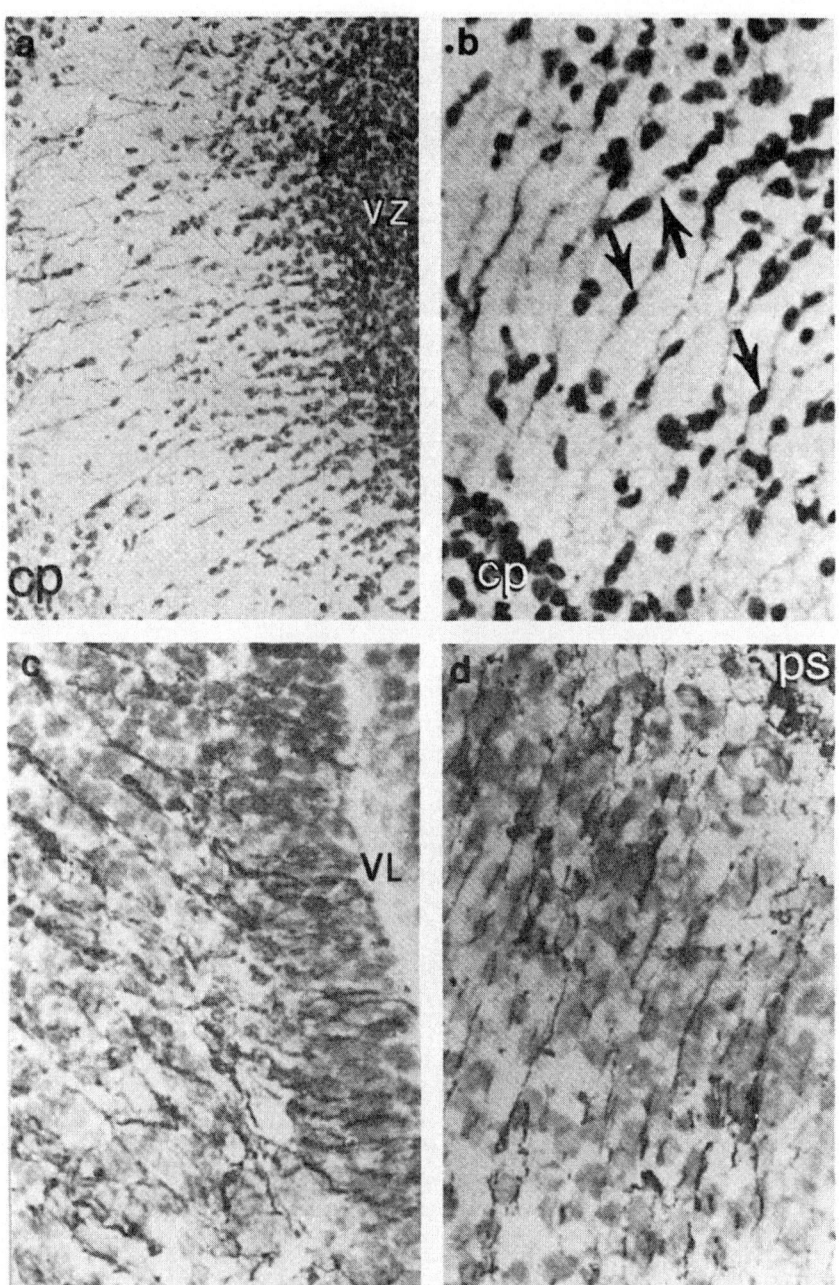

Fig. 14a–d. GFAP immunostaining in the frontal cortex at E18. **a** GFAP-immunoreactive radial fibers are present in the dorsal half of the frontal cortex. *V2*, ventricular germinal zone; *CP*, cortical plate. Hematoxylin counterstaining, ×100. **b** At higher magnification a close association (*arrowheads*) between GFAP-immunoreactive radial fibers and migrating neuroblasts can be observed. *cp*, cortical plate. Hematoxylin counterstaining, ×500. **c, d** GFAP-immunoreactive fibers span the full width of the developing telencephalic wall from the ventricular lumen **c** to the pial surface **d**. *VL*, ventricular lumen; *ps*, pial surface. Hematoxylin counterstaining, ×520

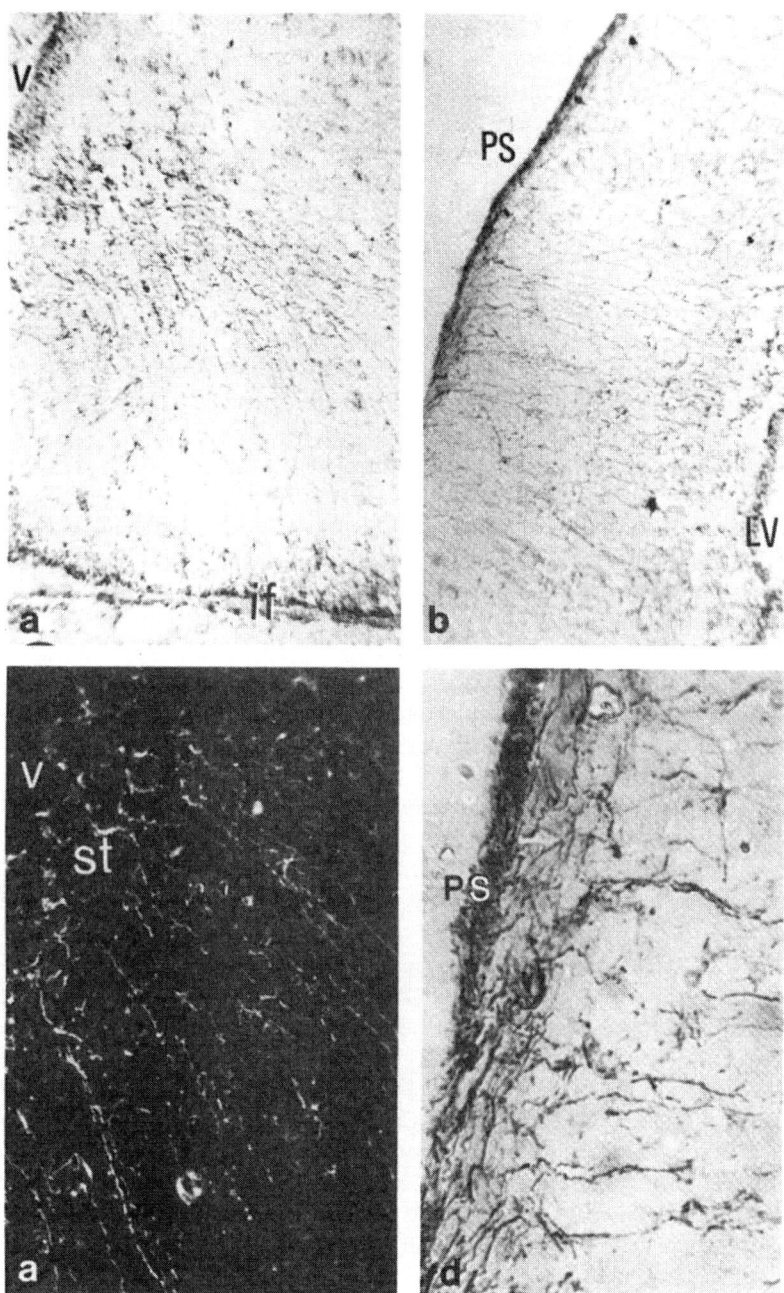

Fig. 15a–d. GFAP immunostaining of the early postnatal telencephalon. **a** GFAP-immunoreactive radial fibers in the frontal cingulate cortex on P4. Fibers span the telencephalic wall from the lateral ventricle (*V*) to the pial surface of the interhemispheric fissure (*if*). × 100. **b** GFAP-immunoreactive radial fibers in the parietal cortex on P4. Fibers span the full width of the telencephalic wall from the lateral ventricle (*LV*) to the pial surface (*PS*). × 100. **c** GFAP-immunoreactive radial fibers originating on P7 in a fan-like fashion at the ventrolateral angle of the lateral ventricle (*V*) from around the stria terminalis (st). Darkfield. × 250. **d** GFAP-immunoreactive radial fibers in the ventral part of the parietal cortex on P14. Note the terminal ramifications under the pial surface (*PS*). × 250

distinguished was the ventral aspect of the brain, i.e., the presumptive pyriform, periamygdaloid, and entorhinal cortices: the cortical plate extended into the entorhinal cortex between E17 and E18, and to the most medial limit of the ventral cortex (the fissura choroidea) by E20. Apart from this region, where radial fibers were only rarely encountered, the ventrodorsal gradient of development of most of the cortical plate clearly contrasts with the dorsoventral gradient of appearance of GFAP-positive radial fibers, the former preceding the latter by up to 7 days, for example, in the lateral and ventral regions. Even where the association between the migrating neuroblasts and radial glia could clearly be demonstrated (E18, frontal cortex), the cortical plate had been present from E16 and was already about 20 cells thick before GFAP-containing fibers became apparent. Within the period studied GFAP-immunoreactive radial fibers were seldom observed in the occipital cortex and in the limbic cortical regions.

4.2.2 Hippocampus and Dentate Gyrus

As mentioned above, the hippocampus was the first area to show GFAP-stained radial fibers, from E16 onwards. Radial stripes of GFAP-positive fibers aligned vertically to the ventricular surface were observed to traverse the hippocampus and dentate gyrus (Fig. 16a–d). These fibers persisted until P9, at which age they disappeared from Ammon's horn. After P14 radial stripes in the fimbria were no longer conspicuous, although radial glial fibers in the dentate gyrus remained strongly stained and persisted even at P30 (Fig. 16c, d), and in the adult (not shown). Glial perikarya in the dentate gyrus were arranged in a single row under the granular layer (Fig. 16d). As in the neocortex, migration of young neurons occurred in the hippocampus before GFAP-immunoreactive radial fibers appeared: a lamina of cells continuous with the cortical plate and representing the incipient stratum pyramidale was well defined by E17 in the subiculum and Ammon's horn.

4.2.3 Diencephalon

GFAP-containing tanycytes were first observed at E17 in the median eminence (Fig. 17). Tanycyte cell bodies were located in the ventricular cell layer of the floor of the third ventricle. Their processes either reached the basal brain surface or terminated in the median eminence. In other parts of the third ventricle GFAP-positive tanycytes were not encountered prenatally. However, after birth GFAP-positive tanycytes were seen to appear gradually throughout the whole lining of the third ventricle, so that by P30 they were present along the whole central diencephalon (Figs. 18, 19a, b, 20a). These cells could still be observed in full grown adult mice. With increasing age staining intensified and tanycyte processes appeared to be thicker than in embryonic stages. GFAP-immunoreactive processes thinned out gradually in the periventricular neuropil. In some cases they could be traced to form contacts on blood vessels, although others either did not make such contacts or could not be followed in the plane of the section.

Similar GFAP-positive tanycytes were seen in the third ventricle of adult rats (data not shown). However, in contrast to the mouse, they were confined to the ventral half of the ventricle, around the infundibular recess.

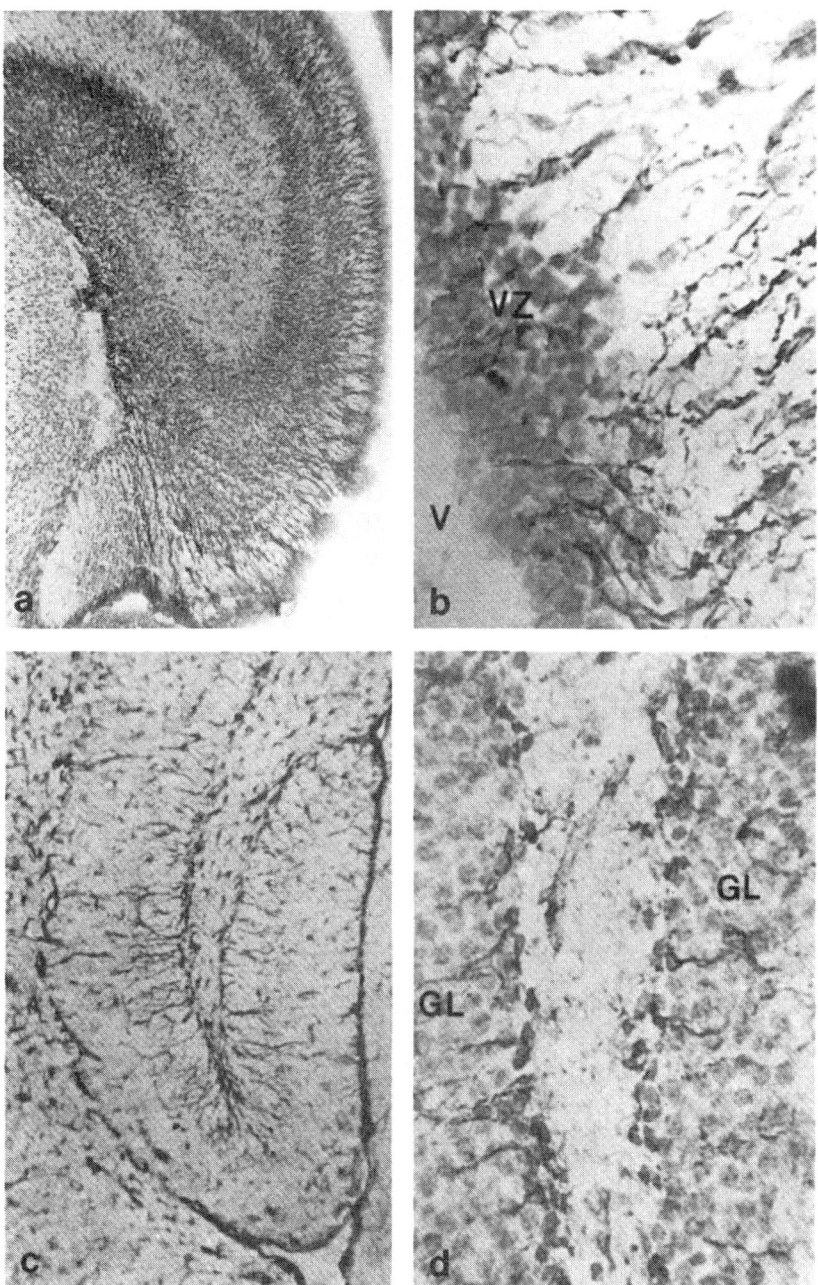

Fig. 16a–d. GFAP immunoreactivity in the hippocampus and dentate gyrus. **a** Low-power survey micrograph of the GFAP-immunostained hippocampus and dentate gyrus. On E18 conspicuous immunoreactive stripes are seen in the fimbria. Hematoxylin counterstaining, × 100. **b** Higher magnification of hippocampal GFAP-immunoreactive fibers visible to reach the ventricular lumen (*V*). *VZ*, ventricular germinal zone. Hematoxylin counterstaining, × 500. **c** At P30 radial fibers of strong GFAP immunopositivity are present in the dentate gyrus. × 100. **d** GFAP immunoreactivity in the dentate gyrus at P30. The position of immunoreactive perikarya in a row under the granular layer (*GL*) is evident. The origination of radial fibers from immunostained perikarya is clearly visible. Hematoxylin counterstaining, × 500

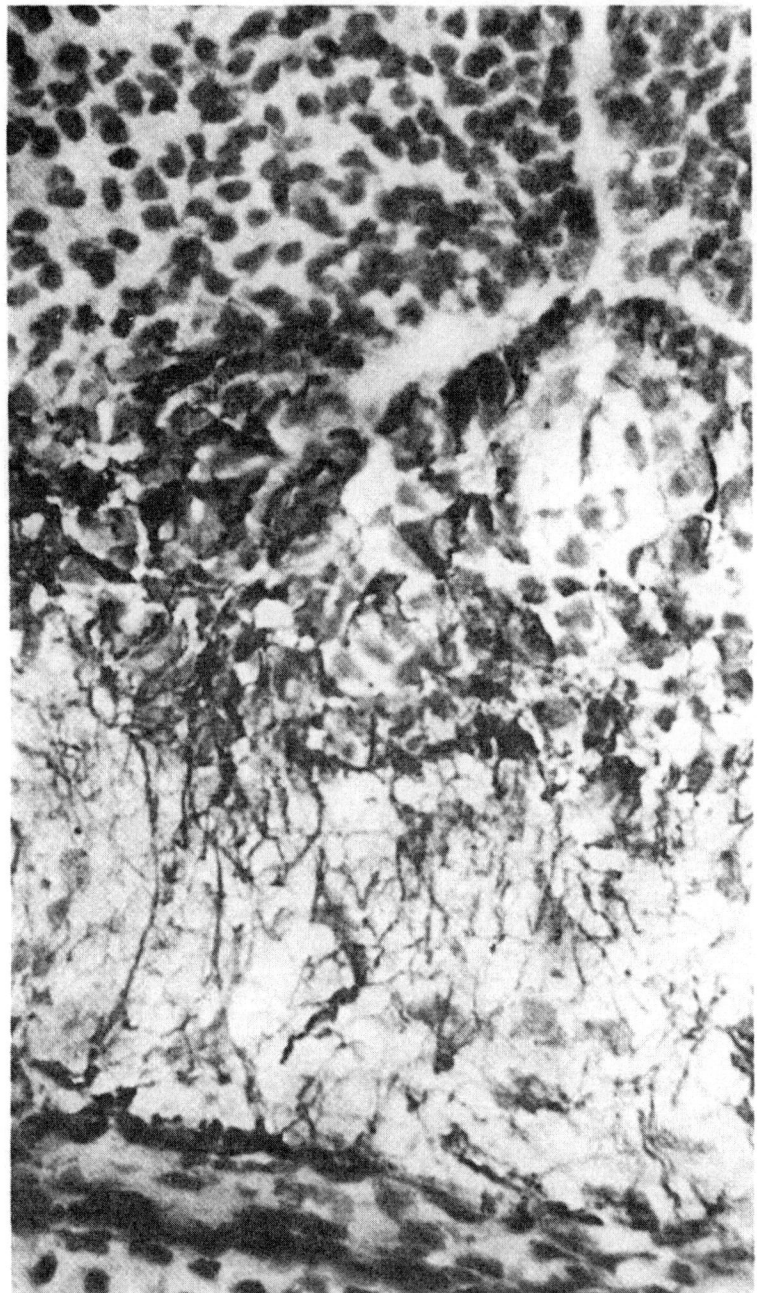

Fig. 17. GFAP reaction in the median eminence of the 17-day-old mouse embryo. Immunoreactive tanycyte cell bodies are located in the ventricular cell-layer of the floor of the third ventricle. Their processes either reach the basal brain surface or terminate in the median eminence. Hematoxylin counterstaining, × 850

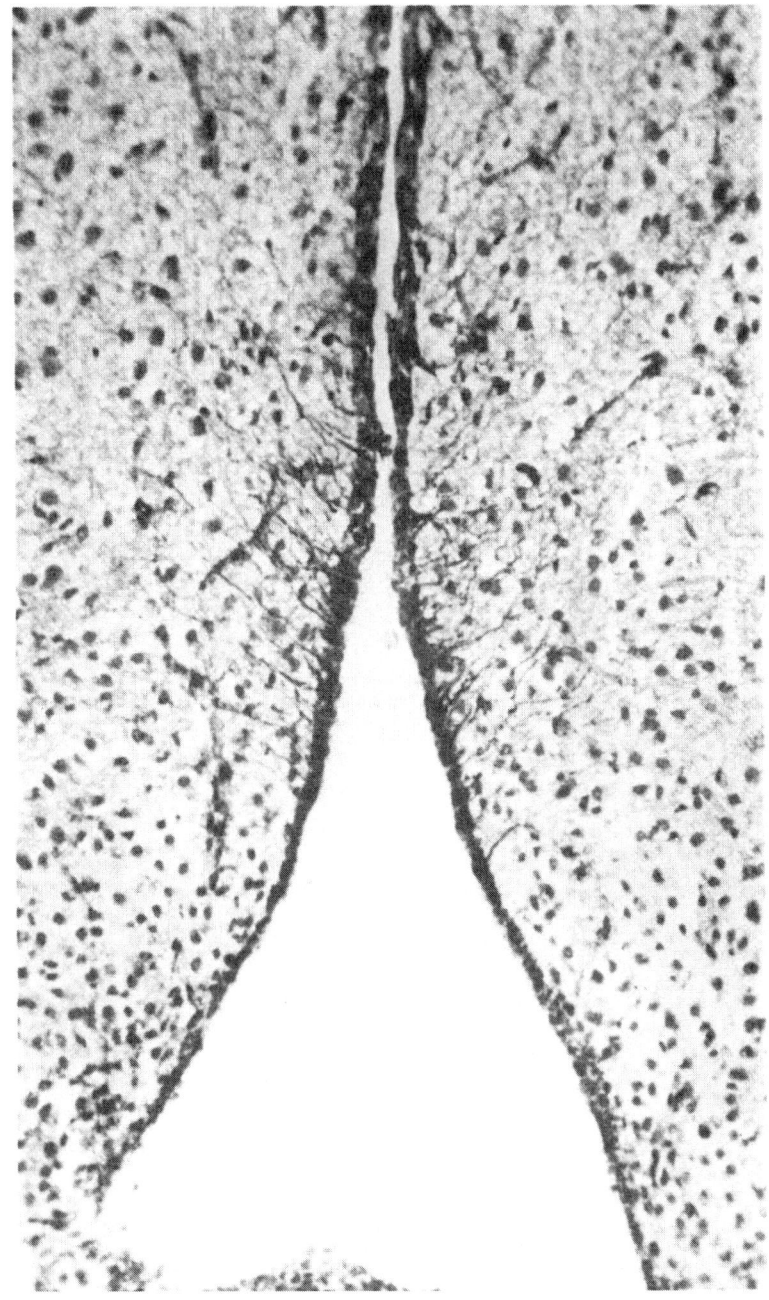

Fig. 18. GFAP-immunoreactive tanycytes in the basal part of the third ventricle of the adult mouse. Hematoxylin counterstaining, × 250

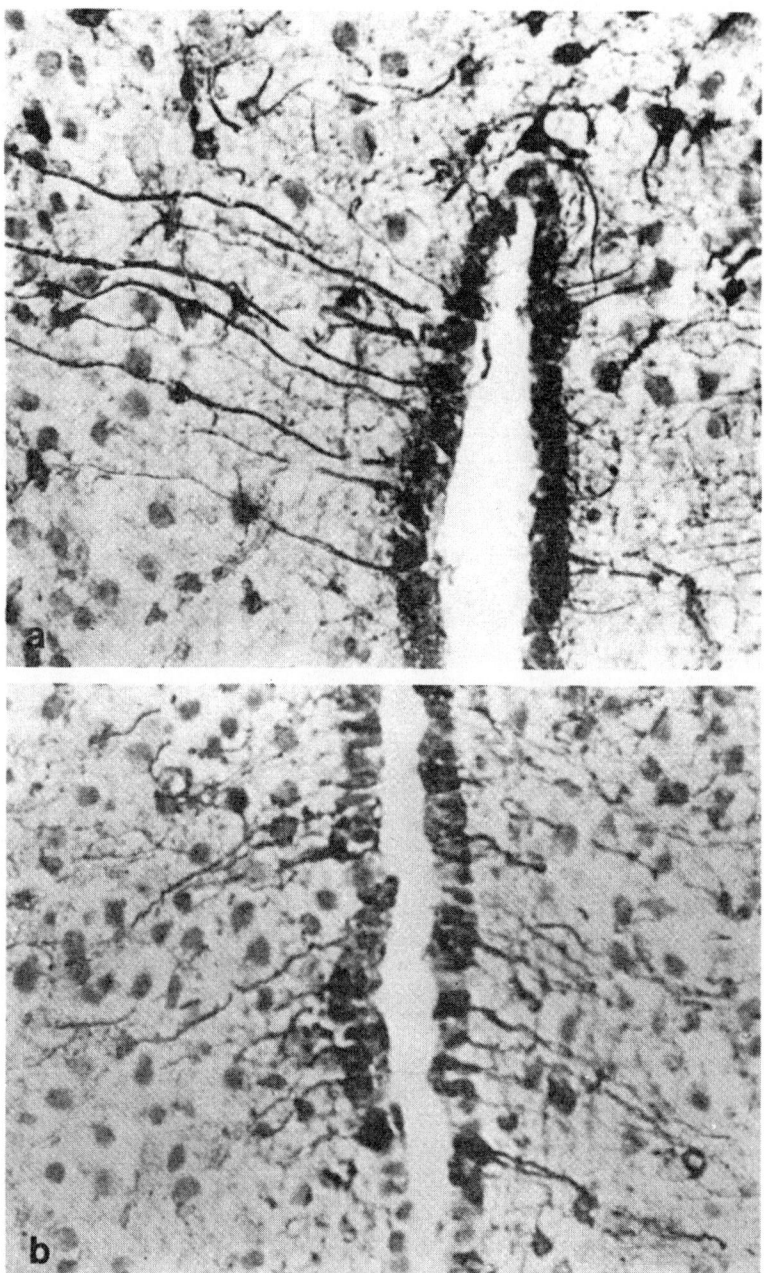

Fig. 19a, b. GFAP-immunoreactive tanycytes around the third ventricle of the adult mouse. Note that GFAP reaction in the tanycyte processes is stronger than in the immature mouse. **a** GFAP-immunoreactive tanycytes at the upper part of the ventricle in the anterior hypothalamus. Hematoxylin counterstaining, × 750. **b** GFAP-immunoreactive tanycytes of the middle portion of the ventricle in the anterior hypothalamus. Hematoxylin counterstaining, × 750

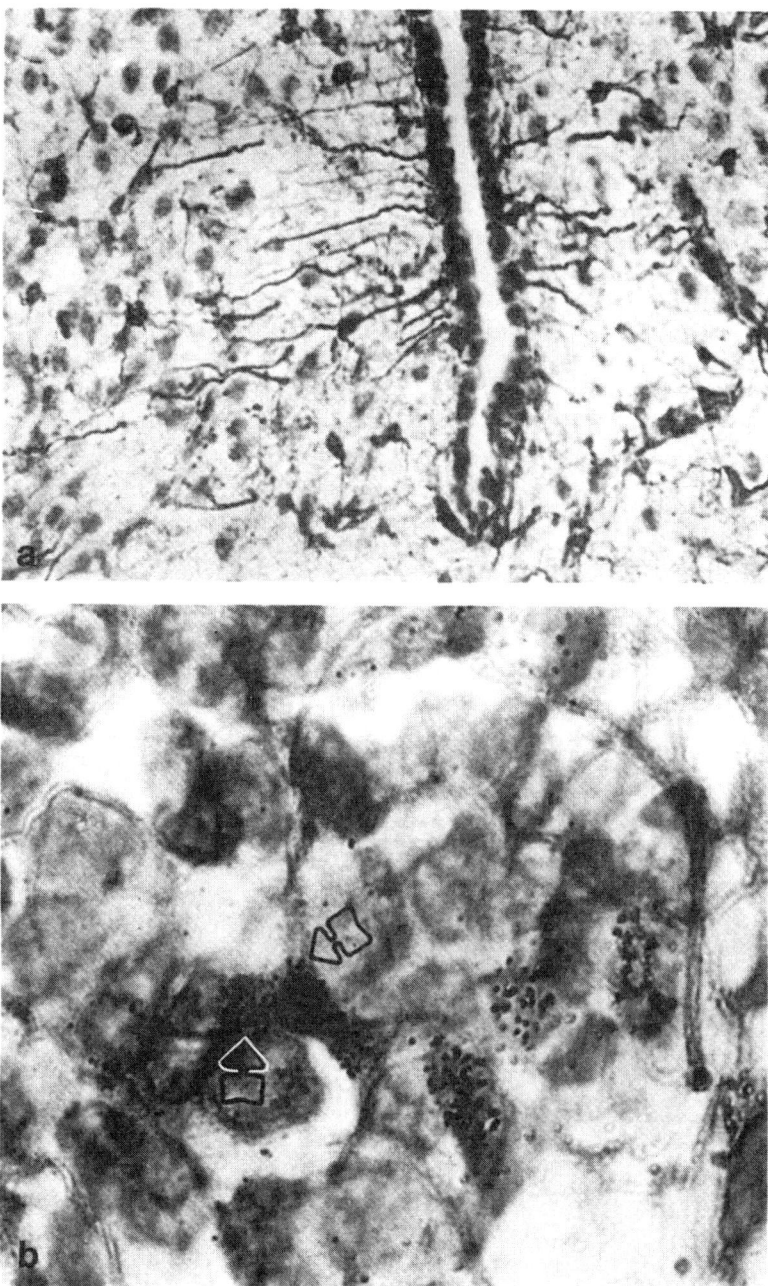

Fig. 20. a GFAP-immunoreactive tanycytes of the basal part of the ventricle in the anterior hypothalamus. Hematoxylin counterstaining, × 750. **b** GFAP staining combined with [³H]thymidine autoradiography in the subventricular zone of the lateral ventricle on E19. A GFAP-immunoreactive cell (*arrows*) with well-developed GFAP-staining processes has a markedly labeled nucleus. The GFAP-positive labeled cell can be distinguished from proliferating neuroblasts. Hematoxylin counterstaining, × 1800

GFAP immunoreactivity was not seen in the ependymal cells proper of developing or adult animals of either species.

4.3 Silver Impregnation of the Forebrain Radial Glia

The capricious nature of the Golgi method precluded a systematic study of the development of radial fibers in different cortical areas, although impregnated fibers were often detectable earlier than GFAP-stained fibers in the same region. For example, Golgi-impregnated radial fibers were seen at E16 in the dorsal telencephalic wall around the lateral ventricle (Fig. 21) while GFAP staining was not demonstrated here until E19. The impregnated fibers were thick and varicose. The perikarya of radial fibers were situated 20–25 μm from the ventricular surface. The cells were bipolar with a short, bulky process extending to the ventricular lumen and a short, slender one directed toward the pial surface where it had a brush-like terminal arborization (Fig. 21).

One day after birth a system of abundant radial fibers could be impregnated in many parts of the cortex (Fig. 22a). Fibers were seen running in the developing subcortical white matter parallel to the surface before curving radially to traverse the cortex (Fig. 22b). Radial fibers of this age were invariably varicose but much thinner than those found at embryonic stages. Club-shaped endfeet under the pia were particularly pronounced from P7 (Fig. 22d). Ramifications except for the typical brush-like terminal arborizations (Fig. 22c, d) were rarely encountered. Fibers of similar appearance were detected up to P12, but at later stages the demonstration of the radial glia was obscured by the predominant impregnation of neuronal elements.

In the hippocampus we failed to demonstrate, with Golgi impregnation, fibers corresponding to the GFAP-positive radial fibers observed in pre- and postnatal mice.

4.4 Comments

The present study demonstrates that radial fibers in the developing cortex contain an astrocyte-specific protein and can thus be clearly regarded as an astroglial cell type. These results in the mouse are consistent with previous reports on the presence of such cells in other species (Antanitus et al. 1976; Choi and Lapham 1978; Levitt and Rakic 1980). We detected the radial glia with Golgi impregnation a little earlier than with GFAP immunocytochemistry. However, satisfactory impregnations were not obtained after P12, although in some cortical areas GFAP-positive radial fibers clearly persisted to P20–30 and even into adulthood. This difference between the two techniques, which arises from the still unexplained selectivity of the Golgi method, was most marked in the hippocampus where, in contrast to GFAP staining, the impregnated radial glia were not detectable at all. In general, it would appear that silver impregnation showed a preference for glial cells in the perinatal period and for neurons later.

The analysis of the development of the radial glia in different regions of the cerebral cortex has demonstrated a lack of correlation between the early waves of migration of young neurons to form the cortical plate and the appearance of GFAP-immunoreactive fibers. Most strikingly, the gradient of develop-

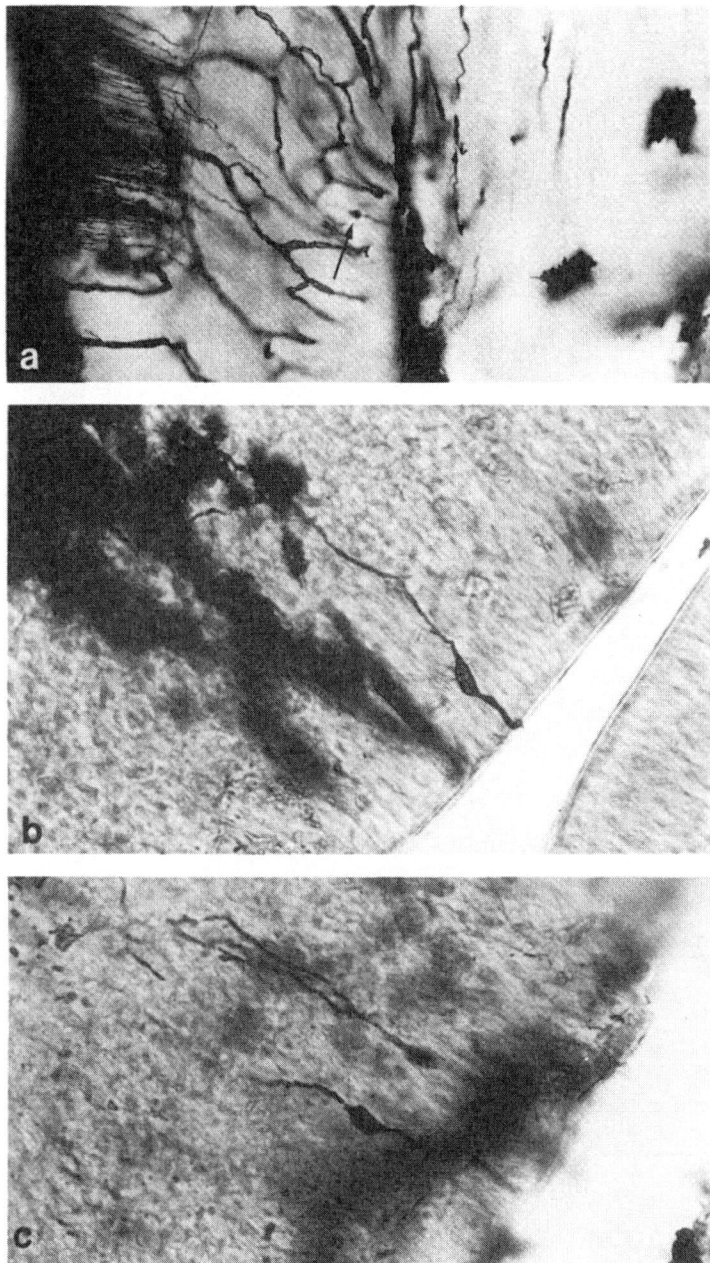

Fig. 21 a–c. Silver-impregnated radial glia at E16 in the telencephalic wall. **a** Low-power view of an impregnated radial glial cell (*arrow*), which is bipolar with a short, centrally directed process and a long, slender one running up to the pial surface, where it ends with a fork- or brush-like terminal arbor, × 250. **b, c** The position of the perikaryon relative to the ventricle is clearly seen. In **b** the bulby central process can be observed to contact the ventricular lumen

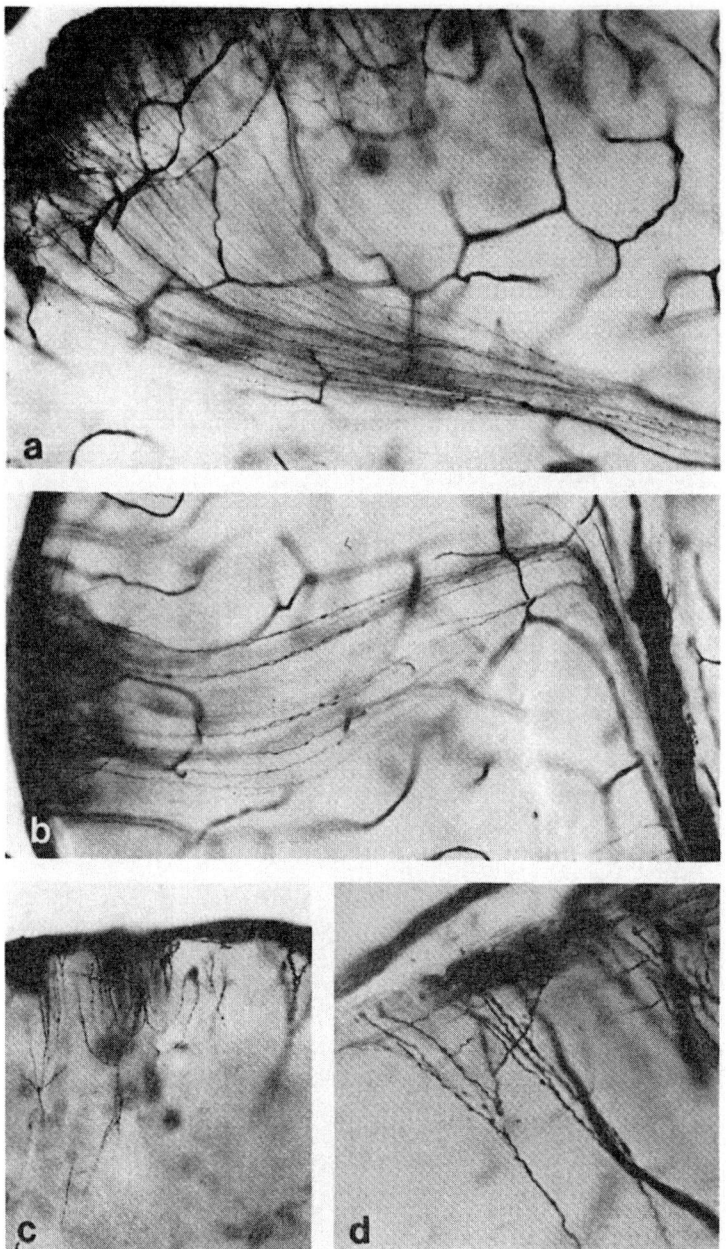

Fig. 22a–d. Silver impregnation of radial fibers in the early postnatal period. **a** On P1 a system of abundant finely varicose radial fibers is impregnated throughout the parietal cortex, × 250. **b** Radial fibers of the parietal cortex run first in the telencephalic white matter parallel to the surface, and then turn with a sharp bend to the perpendicular and traverse the telencephalic wall, × 300. **c** Typical brush-like termination of radial glial fibers under the pia. P7, × 300. **d** Terminal ramifications of radial fibers with club-shaped end feet under the pial surface. P7, × 500

ment of the cortical plate proceeded in an opposite direction (ventrodorsally) to that of the immunoreactive radial glia (dorsoventrally). It must be stressed here that the discrepancy lies between the patterns of development rather than with the earliest age of detectable immunoreactivity: the technical limitations of the latter are obvious, and the early radial glia possessing long processes could be present in some areas, despite the failure to detect it by the Golgi or GFAP techniques. Moreover, Levitt et al. (1981) have recently described in the fetal monkey occipital cortex glial precursor cells which showed GFAP immunoreactivity while undergoing mitosis, although they appeared to lack long processes. This question warrants further study, especially in view of recent observations suggesting that certain antigens may be expressed before GFAP in immature astrocytes and/or astrocyte precursor cells (Bartlett et al. 1981; Chiu et al. 1981; Dahl et al. 1981). Other antigens have been reported which appear transiently in astrocytes and at different times, depending on the region (Lagenaur et al. 1980), although they were not exclusive to this cell type. However, even with these qualifications, it is striking that the sequence of cortical plate formation and that of the detectable radial glia proceeds in opposite directions, and that in certain regions, particularly those at the base of the brain, we could detect virtually no glial fibers at any stage of development. Since we have examined the latter areas at various stages from E14 to adult, it seems rather unlikely that the lack of GFAP immunoreactivity could be ascribed to immaturity of the fibers.

The apparent absence of the radial glia in some areas and their pattern of development in the mouse argue against a universal applicability of the hypothesis of glial guidance of neuronal migration, and in this context it is of interest that a degree of cortical complexity and lamination can be obtained in reaggregating cells without any apparent involvement of the radial glia (Garber et al. 1980). Our results on the formation of the cortical plate in different brain regions agree with the time course of the birthdays of mouse cortical neurons described by Angevine and Sidman (1962) and Caviness et al. (1981) and with other results from the rat (Raedler and Raedler 1978; Altman 1969). The last three studies estimate that in the early stages of cortical plate formation the time of migration of neurons from the ventricular zone to their final destination is about 2 days. Thus in order for the glial fibers to act as guides for the migration of young neurons to the cortical plate they would need to be present in the mouse lateral cortex, for example, from about E13. Our data show that GFAP-positive fibers are not found here until about P4, a difference in timing which seems to be too long to be accounted for by a delayed appearance of GFAP immunoreactivity in "immature" glia, which nevertheless have fully developed radial fibers and are present at the appropriate time.

The foregoing discussion argues against the radial glia being a prerequisite for the formation of the cortical plate in the mouse, although Rakic and his co-workers have clearly demonstrated an association between migrating neurons and the radial glia in various regions of the primate brain (Rakic 1971a, b; Sidman and Rakic 1973; Nowakowsky and Rakic 1979; Levitt and Rakic 1980). While we also recognized the same association during the later stages of prenatal brain development in the mouse (e.g., Fig. 13b), there were notable differences between our results and those of Rakic. For example, the radial glia was sparse in the parieto-occipital region of the mouse brain, the area most extensively studied in his work on the monkey. These discrepancies may well be accounted for on a phylogenetic basis. Levitt and Rakic (1980) point out that in the monkey migrating neurons may have to travel relatively long distances and thus the radial glia may provide assistance in traversing a complicated terrain across orthogonal tracts before they reach their destination. In rodents the distances involved are much smaller and the radial glia appear to be less extensive, particularly in the early stages of formation of the cortical plate. The possibility of glial guidance operating later rather than in the very early stages of development in some species has also been suggested by other workers (Shoukimas and Hinds 1978). We have noted interspecies differences in the distribution and postnatal persistence of the radial glia (Bascó et al. 1981), and some of the apparent discrepancies in the literature regarding the presence of absence of the radial glia may be accounted for by differences in the ages and species studied.

The radial glia of the hippocampal dentate gyrus are reminiscent of the cerebellar Bergmann glia, particularly with respect to their persistence in the adult and the orientation of the perikarya in a single row. It may be relevant that in both regions interneurons proliferate and migrate for some time postnatally (Altman 1969; Bayer and Altman 1974; Jacobson 1978).

Previous authors who stained the ventricular lining for GFAP reported that the ependyma is GFAP-negative, but while Ludwin et al. (1976) did not mention tanycytes, other workers briefly noted occasional positive cells which resembled tanycytes (Deck et al. 1978; Velasco et al. 1980). In a more detailed study Roessmann et al. (1980) described GFAP immunoreactivity of both ependymal cells and tanycytes in the human third ventricle. However, this was transient, being present in the fetus but not in the adult, in contrast to our results from the rodent brain which showed that GFAP immunoreactivity persists in adult tanycytes. This discrepancy between species is probably a reflection of wider phylogenetic differences in the distribution and presumably function of tanycytes. In lower vertebrates, for example, they are widely spread throughout the ventricular system (Horstmann 1954; Millhouse 1972). The demonstration of GFAP in the radial glia of the developing forebrain (Bignami and Dahl 1974b; Choi and Lapham 1978), in the cerebellar Bergmann glia (Bignami and Dahl 1973, 1974c), and now in tanycytes shows that tanycytes belong to a family of cells which share certain immunospecificities and which comprise a wider variety of cell types than classical astrocytes.

It is also of interest that GFAP-immunoreactive tanycytes were first seen in the late embryo around the infundibular recess, while in the adult mouse (but not the rat) GFAP staining revealed a fairly uniformly distributed tanycyte system around the whole third ventricle. GFAP immunoreactivity develops at a certain maturational stage of the cell, and our findings in the embryo may thus show fewer tanycytes than actually present. On the other hand in the adult brain glial elements are known to be difficult to detect by the silver impregnation techniques which have hitherto been employed in many investigations. There is also a regional selectivity of this method, all of which may lead to reduced demonstrability of tanycytes along certain parts of the ventricle. Thus while immunocytochemical data from the embryo may give an underestimate of the tanycyte population, in the adult the GFAP reaction probably gives a better indication of their number and distribution than silver staining.

44

Thus defined in terms of GFAP immunoreactivity, the astroglia comprises a wide variety of cell types, some of the cells retaining their processes into adult life while others may be transformed into conventional astrocytes (Choi and Lapham 1978; Schmechel and Rakic 1979a). The view that the radial glia gives rise to conventional astrocytes (Ramón y Cajal 1911) is consistent with the observation that it appears to be capable of proliferation while having processes extended to the brain surface (Bascó et al. 1977; Hajós et al. 1981), as also shown in earlier studies on the cerebellar Bergmann glia (Bascó et al. 1977). In this context it is of interest that the cortex surrounding the middle segment of the stria terminalis contained immunoreactive radial fibers even at P30, and this site has been described as retaining postnatal proliferative capacity for a long time (Smart 1961; Smart and Leblond 1961; Schimrigk 1966; Fleischauer 1970; Schlessinger et al. 1975). Alternatively, the prolonged presence of the radial glia in certain regions of the brain but not in others may be interpreted as a phylogenetic remnant of the more widespread system of radial fibers found in lower vertebrates (Horstmann 1954) or as a persistent direct connection between ventricular and pial surfaces.

5 Demonstration of Proliferative Capacity of the GFAP-Immunoreactive Radial Glia

5.1 Introductory Remarks

Our earlier findings in the cerebellum (Bascó et al. 1977) and now in the forebrain suggest that the proliferation of the glia involves considerably differentiated cells as judged by morphological criteria such as position, ramification pattern, and cytoplasmic and nuclear structure. A spectacular example of this phenomenon was presented in Sect. 3.2, where the cerebellar Bergmann glia were shown to undergo divisions while having a well-developed fiber system extended to the pial surface.

The retained proliferative capacity of the glia over a long postnatal period has been established in various regions of the brain (Allen 1912; Altman 1962, 1966a; Hommes and Leblond 1967; Dalton et al. 1968; Gilmore 1971; Hinds 1968a, b; Korr et al. 1973; Lewis 1968a, b; Sturrock 1974a).

Evidence that the GFAP-immunoreactive radial glia also share this property has so far been purely circumstantial, based on similarities between the proliferating glia in general and the radial glia. To obtain direct proof for the proliferation of the radial glia [³H]thymidine autoradiography and GFAP immunostaining were carried out successively in the same preparations. This type of combination has not been used so far and was made possible by the remarkable stability of the GFAP reaction product, enabling a subsequent autoradiographic procedure to be performed in the same section.

As far as this combined technique is concerned it has to be noted that under the light microscope the yellowish-brown color of the GFAP staining is clearly distinguishable from the black silver grains; and by the use of a light counterstaining of the nuclei with hematoxylin giving a pale blue tone to the background, a really impressive and colorful picture can be obtained [see the color picture in Hajós et al. (1981)]. Therefore, to give more credit to the forthcoming description illustrated in black-and-white, the reader is referred to the illustration in the paper of Hajós et al. (1981).

5.2 [³H]Thymidine Uptake into the GFAP-Immunopositive Radial Glia

Silver grains due to pulse-labeling with [³H]thymidine were observed over the nuclei of several GFAP-immunopositive cells situated in the subventricular zone of the telencephalon (Fig. 20b). These cells had well-developed processes also showing the GFAP reaction. They could be safely distinguished from the surrounding proliferative neuroblasts. Connection of the proliferating radial glia with the ventricular lumen was also seen. Immunostained, isotope-labeled radial glia were observed from E16 to P5–6.

5.3 Comments

Findings with the combined application of GFAP immunocytochemistry and [^{3}H]thymidine autoradiography indicate that the biochemically mature astroglia (as judged by their content of a specific protein) is capable of proliferation. The retained proliferative capacity of the astroglia has been observed elsewhere under normal (Allen 1912; Altman 1962, 1966a; Hommes and Leblond 1967; Sjöstrand 1965) and experimental (Altman and Das 1964; Cammermeyer 1963; Kreutzberg 1966; Cavanagh 1970; Chow and Dewson 1966; Diamond et al. 1964; Kuhlenkampf 1952; Murray 1968; Sjöstrand 1965; Watson 1974) conditions.

The question arises as to whether differentiation has been arrested at an early stage, or whether mature astrocytes undergo divisions (cf. Jacobson 1974) and we have investigated this by means of the combined application of immunocytochemistry and autoradiography. It was found that the presence of GFAP in the cell body of an elaborate GFAP-containing fiber system is not incompatible with the replication of the cell. Similar conclusions were arrived at in earlier histological studies (Bascó et al. 1977; Bascó 1981). Therefore, there is no need to suppose the presence of dormant undifferentiated cells to explain astroglial proliferation at later stages.

6 Transport of Material by Glial Processes

6.1 Introductory Remarks

In the foregoing we pointed out the developmental, morphological, and immun-ocytochemical similarities between cells thought so far to belong to distinct classes. As far as functional similarities of these cell types are concerned no common features are mentioned in the literature. Most glial functions known to date, such as K^+ transport (Kuffler and Nicholls 1966; Kuffler 1967; Orkand 1977), uptake of amino acid transmitters (Henn and Hamberger 1971; Hösli and Hösli 1976; Schousböe et al. 1977), and compartmentation of energy metab-olism (Hamberger and Sellström 1975) have not been comparatively investigated in all cell types related to the glia. As a matter of fact, such comparative investi-gations of functions of various glial populations would require further sophisti-cation in bulk cell separation techniques, most of which, at least for the time being, suffer from the drawback of uncertainty and a high degree of cross contamination.

There appears to be, however, a function the demonstration of which is feasible in tissue sections, thus enabling comparative studies to be carried out, and which has already been verified in tanycytes, a cell class clearly shown to contain an astrocyte-specific protein in common to the forebrain and the cerebellar radial glia. The long processes of tanycytes terminating either at the pial surface of the base of the brain or around blood vessels suggested to the investigators a function of transporting materials between various extracellular spaces (Löfgren 1959; Rodriguez 1972; Scott et al. 1974; Léránth and Schiebler 1974; Wagner and Pilgrim 1974; Nakai and Naito 1975). This function, which is more than probable in tanycytes, has also been proposed for other types of glia having long processes, in particular for the radial glia (Oksche 1968).

The introduction of the HRP tracer technique (Kristensson et al. 1971; La Vail and La Vail 1972) opened new perspectives in the study of neuronal net-works but has not been fully exploited for glial research. Recently, Ivy and Killackey (1978) reported on a transient cell population — glial in their opinion — of the immature rat brain that could be labeled by the retrograde transport of HRP deposited in the upper layers of the parietal cortex. However, due to the early presence and continuous ingrowth of the cortical afferents (Wise and Jones 1976, 1978) known to be capable of HRP uptake, this method of tracer administration can be used only during a relatively short postnatal period for the analysis of the glia. To solve this problem we have deposited HRP onto the pial surface without interrupting the external limitans membrane formed by glial end feet. We followed the assumption that if glial processes have in general the capability of material transport similar to that attributed to tanycyte processes, the radial glia of the forebrain and the cerebellar Bergmann glia would also be expected to concentrate the transported tracer in their cell bodies.

6.2 Transport of HRP by the Forebrain Radial Glia and Cerebellar Bergmann Glia

6.2.1 Transport in the Forebrain

Horse radish peroxidase was deposited at various sites of the exposed pial surface of the cerebral cortex of infant and adult mice. The HRP depot, owing to the way of deposition, appeared to diffuse along the subarachnoid space. In the present context this proved to be an advantage rather than a disadvantage as a larger surface area covered by glial end feet was exposed to the tracer. However, as indicated by serial sections, the site of the original deposit could still be recognized.

A marked labeling of cell bodies situated in the subventricular zone of the central part of the lateral ventricle was observed in the 4-day-old (P4) animal (Fig. 23a, b). At postnatal day 7 (P7) the territory containing labeled cell bodies was smaller and by P10–12 it became restricted to the area at the ventrolateral corner of the lateral ventricle. Labeled cells of this age were located along the stria terminals (Fig. 24a, b). In the frontal plane it became apparent that they surrounded the stria. Occasionally some radially directed fibers were also labeled. In other preparations the typical asymmetrically bipolar shape of labeled radial glia cells was evident. Under oil immersion the label proved to be of the spherical-granular nature indicative of a retrograde transport (Mesulam 1976). In animals older than 12 days no detectable accumulation of HRP was found in the subventricular zone of the telencephalon.

These findings delineate the ventrolateral ventricular corner as a site of prolonged persistence of glial cells communicating with the pial surface. This is in good agreement with the findings of Fleischhauer (1970).

6.2.2 Transport in the Cerebellum

In the cerebellar cortex the deposition onto the pial surface of HRP resulted in strong labeling of the Bergmann glia in the immature as well as in the mature animals. In particular, the labeling of the adult Bergmann glia was remarkable (Fig. 25a). Accumulated HRP outlined not only the ganglionic layer where Bergmann glia cell bodies are situated, but it often revealed the typical shape of the initial ramifications of the Bergmann fibers in the inner part of the molecular layer (Fig. 25b). Accumulation of the tracer was so massive that it almost solidly filled the cell body and the proximal portion of the radially directed processes and their characteristic irregular appendages.

6.3 Comments

Our HRP studies have shown that the radial glia of the forebrain and the cerebellar Bergmann glia possesses a capacity for transporting material similar

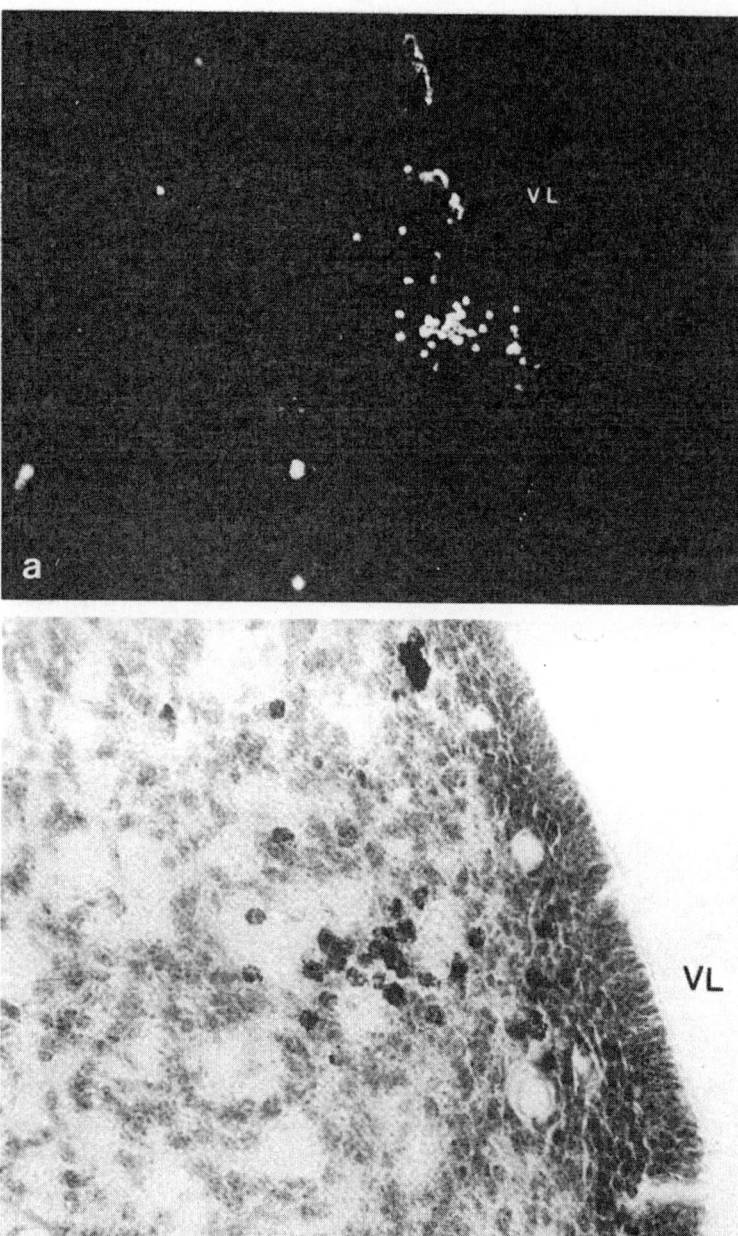

Fig. 23a, b. Uptake of HRP deposited onto the pial surface on P4. **a** Darkfield micrograph of the ventral part of the lateral ventricle. A group of labeled cells are located near the lateral wall of the ventricle. *VL*, lateral ventricle, × 200. **b** Slightly enlarged brightfield micrograph of the area shown in **a**. The position of the group of labeled cells can clearly be assessed relative to the ependymal lining and germinative zone of the ventricle. *VL*, lateral ventricle. Safranin counterstaining, × 300

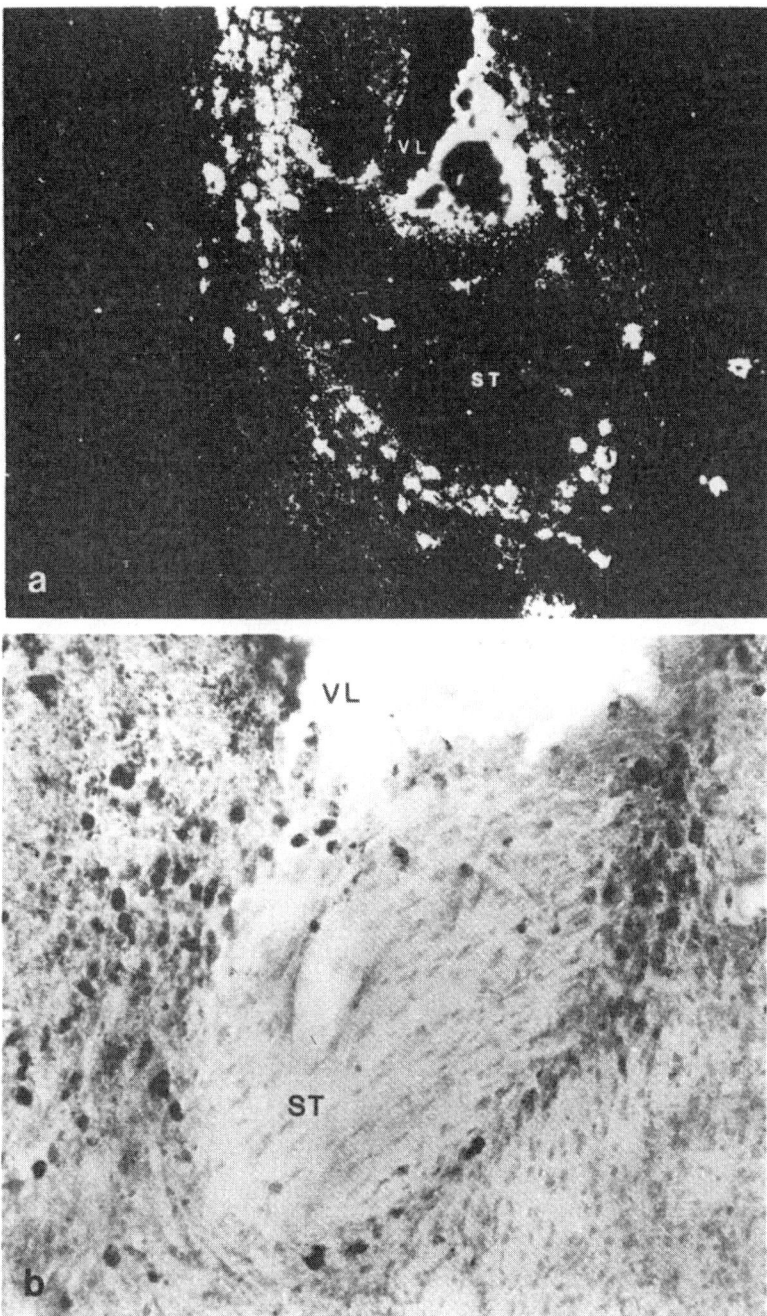

Fig. 24a, b. Uptake of HRP deposited onto the pial surface on P12. **a** Darkfield micrograph of the ventrolateral angle of the lateral ventricle. Labeled cells surround the stria terminalis (*ST*). *VL*, lateral ventricle, × 300. **b** Brightfield micrograph of an area at the ventrolateral ventricular angle similar to that shown in **a**. The position of labeled cells relative to the lateral ventricle (*VL*) and stria terminalis (*ST*) is well documented. Safranin counterstaining, × 300

51

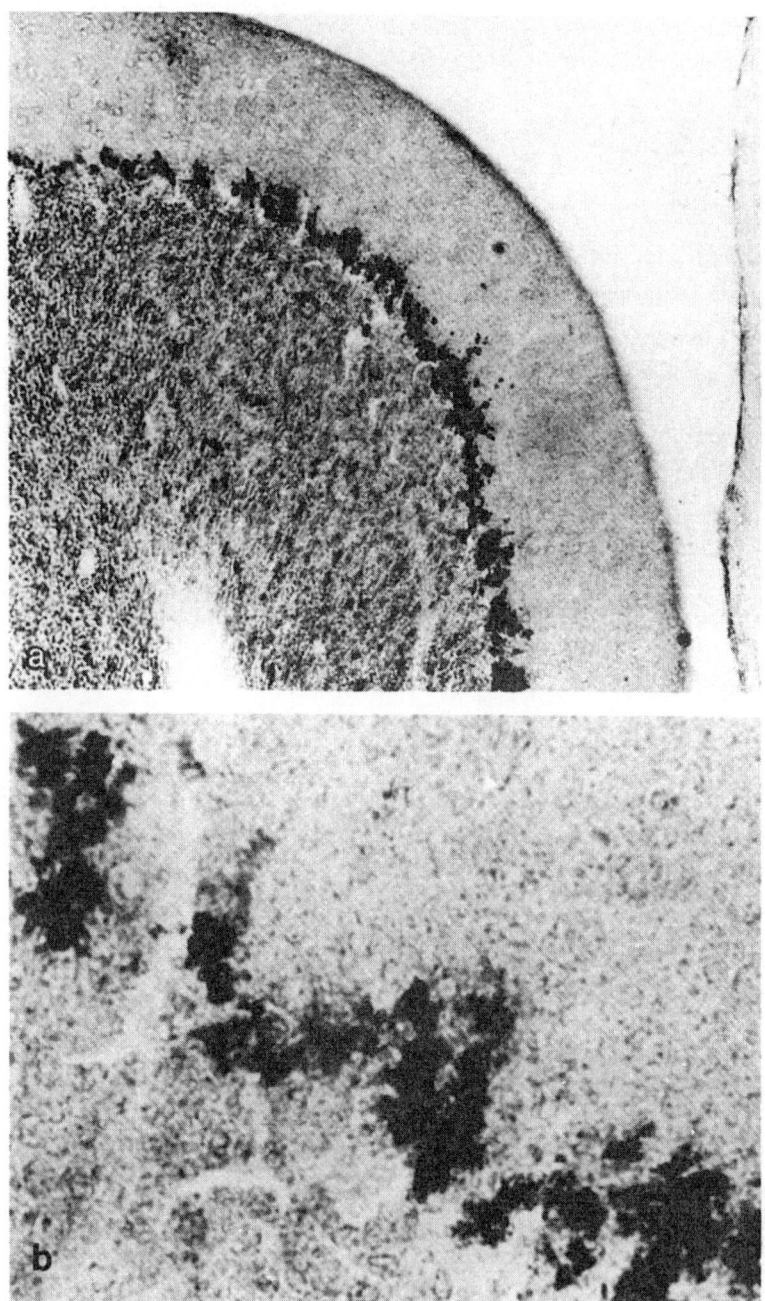

Fig. 25a, b. Transport of HRP from the cerebellar pial surface in the adult mouse. **a** Accumulation of the tracer occurs in the ganglionic layer where the cell bodies of the Bergmann glia are located. Accumulated HRP also outlines the initial portion of Bergmann fibers in the inner molecular layer, × 170. **b** At a higher magnification the characteristic initial arborization of Bergmann fibers can be perceived, × 700

to that demonstrated for diencephalic tanycytes (Rodriguez 1972; Léránth and Schiebler 1974; Scott et al. 1974; Wagner and Pilgrim 1974; Nakai and Naito 1975) and the forebrain radial glia (Chu Wong et al. 1981; Hajós et al. 1982). A further relationship between these cell types has been shown by the immunocytochemical demonstration of their content of an astrocyte-specific protein (Bignami and Dahl 1973, 1974a, b; Roessmann et al. 1980; Levitt and Rakic 1980; Bascó et al. 1981). Thus it appears that the above three cell types are of astrocytic nature and are capable of transporting material along their processes.

This latter phenomenon, however, is merely indicated by the present study as well as in the works quoted. Little is known at present about the parameters of HRP transport along glial fibers. Data available only document the existence of such a mechanism; its directionality, rate of transport, and dependence on neuronal and other environmental factors still have to be elucidated.

As to the direction of transport the present findings prove the existence of a retrograde transport which clearly follows from the experimental strategy employed (i.e., administering the tracer at the end feet region of glial cell types studied). The retrograde nature of the transport seems to be also indicated by the granular nature of the label in the cell body. This according to Mesulam (1976) is characteristic for HRP accumulated by retrograde transport, in contrast to the solid precipitation of the cell body resulting from anterograde transport. However, we have observed the spherical-granular type of precipitation only in the cell bodies of the forebrain radial glia, whereas the cerebellar Bergmann glia showed a solid filling with the tracer. As the present mode of tracer administration (dropping of HRP on the pial surface) rules out the possibility of any kind of anterograde transport, the claim of Mesulam (1976) regarding the diagnostic value of the appearance of perikaryal HRP label has to be viewed critically. It may well be applicable for one or another brain area but its general validity must be questioned.

A further problem is whether transport in the glial processes is uni- or bidirectional. In neuronal processes bidirectional transport is well established [see for review Weiss (1982)]. The work of Léránth and Schiebler (1974) suggests that in the long processes of hypothalamic tanycytes there exists an anterograde transport, forwarding HRP injected to the ventricle toward the hypothalamic blood vessels. There is ample reason to believe that other types of the radial glia also possess the capacity of bidirectional transport.

Even less is known of the rate of transport in glial processes or about the possible coexistence of rapid and slow transport mechanisms within the same process, a phenomenon quite general in the case of neuronal processes.

What can be said at the present stage of investigations about transport mechanisms in radial glia-related cell processes is that they seem to be similar in their basic properties to transport phenomena in other cells, but the elucidation of their specific properties still requires further experimental efforts.

7 Discussion

7.1 Proliferating Cells at Nongerminal Sites in the Early Postnatal Period

An important fact that emerges from the present results is that between P1 and P14 much if not all of the proliferating cells at nongerminal sites of the brain (i.e., outside the subventricular zones, the granular layer of the dentate gyrus, and the external granular layer of the cerebellar cortex) are related to the embryonic radial glia.

Proliferative cells were shown to belong to a distinct population of macroglial cells different from oligodendroglial cells. The basis of this claim was primarily a set of circumstantial evidence, for example: (a) timing of proliferation, (b) localization and morphological appearance of proliferating cells, and (c) their relationship to neurons.

(a) Between P1 and P14 a wave of cell proliferation and migration occurs in the brain which follows the mass multiplication of neurons in the late embryonic period. This postnatal wave of proliferation and migration was termed by Lorente de Nó (1933) as the "second migration." By P14 this "second migration" declines, and then a sudden increase in proliferative activity is observed after the second postnatal week. This latter could clearly be related to the oligodendroglia which undergo rapid multiplication in connection with the onset of myelination of large fiber tracts.

(b) Proliferating cells were often seen in the neuropil of the forebrain. Their mitotic figures were large, pale, and irregularly outlined. The origination of processes could frequently be observed. In the cerebellum the situation was more complicated as in the external granular layer neurogenesis continues on a large scale in the first three postnatal weeks, and since the last century there have been claims (Bellongi and Stefani 1889; Obersteiner 1983; Schaper 1894; Fujita 1967) that this layer also gives rise to the Bergmann glia. Our present results and the double-label autoradiographic studies of Moskovkin et al. (1978) contradict the above claim since they clearly show that the isotope-incorporating cells of the internal granular layer synthesize DNA inside the layer and there is no sign whatsoever of labeled internal granular layer cells being acquired from outside the layer. Long-interval double-labelings (Moskovkin et al. 1978) were also indicative of repeated mitoses within the internal granular layer. These cells, prior to their proliferation, migrate presumably to the internal granular layer prenatally together with the Purkinje and Golgi neurons as it was suggested by the first descriptions of cerebellar histogenesis (Athias 1887; Ramón y Cajal 1890). The primary subventricular origin is strongly supported by the observation of Del Cerro and Swarz (1976), Swarz (1976), and Swarz and Del Cerro (1977), who with electron microscopy have observed the presence of Bergmann-fibers prenatally, well before the formation of the external granular layer. Moreover, the work of Seress (1978) demonstrated an insensitivity to thyroid hormone of proliferating internal granular layer cells in contrast to the high sensitivity of external granular layer cells. All these findings point to the continuous proliferative activity of a cell population within the internal granular layer of the cerebellar cortex from the late prenatal until the end of the early postnatal period.

(c) Proliferative cells throughout the forebrain were often seen in a satellite position to large neuronal perikarya. In the dentate gyrus their characteristic subgranular localization was observed, while in the cerebellum they gradually occupied a position in the ganglionic layer, around the Purkinje cells.

More direct observations were made with GFAP immunocytochemistry, which provided evidence for the identity of these postnatally proliferating cells with the elements of the radial glia meshwork of the brain that degrades gradually during this period.

Comparing the regional distribution of labeled cells with the time course of mitotic activity in the period between P1 and P14 and with the disappearance of the radial glia, it appears that these events follow each other along similar paths. During maturation the rostrocaudal gradient of mitotic peaks parallels nicely with a previous disappearance of the radial glia showing the same temporal and regional gradients. It has to be noted that this gradient is also a phylogenetic one. While in mammals most of the original radial glia disappears by adulthood, gradually less of it is lost during development in lower species in a descending phylogenetic order. This trend can be followed down to some invertebrates where an analogue of the radial glia is permanent in the entire central nervous system (Horstmann 1954; Ariëns Kappers et al. 1960). In other words: the lower a species stands in the phylogenetic hierarchy, the more of the radial glia is retained toward the rostral part of its central nervous system.

All these points enable us to state that proliferative cells at nongerminal sites of the immature brain represent the proliferation of the radial glia and derivatives. This glial proliferation follows a well-defined regional program with regular onset, peak, and cessation times coinciding with the gradual reduction of the radial glia in most areas of the brain. This is consistent with the view (cf. Jacobson 1978) that glial cells in any region withdraw from the mitotic cycle over a much longer period than the neurons of the same region.

7.2 Persistence of the Radial Glia

Although in mammals the radial glia is generally referred to as an embryonic transient cell type, there is an area where its prolonged postnatal presence is remarkable. Emanation of GFAP-immunoreactive radial fibers from the ventrolateral angle of the lateral ventricle was seen even after P30. This was the area to which HRP deposited onto the pial surface of the parietal cortex was transported. The labeled cells as well as the origin of emanating GFAP-immunopositive fibers showed a localization along the stria terminalis. It is noteworthy that this is the place where cells are thought to retain their proliferative capacity throughout the animal's life (Smart 1961; Smart and Leblond 1961; Schimrigk 1962; Fleischhauer 1972; Schlessinger 1975) and which, as suggested by experience gained in animal and human neuropathology, is a frequent source of tumor formation (Lewis 1968b; Wollemann 1974). Most of these tumors have proved to be astrocytomas, which is in good agreement with our conclusion that these proliferating cells are of an astrocyte nature.

7.3 Does Postnatal Glial Proliferation Involve Dormant Stem Cells or the Differentiated Glia?

Relevant to possible tumor formation from radial glial elements, a further interesting question of postnatal glial proliferation is whether new cells are formed from dormant cells whose proliferation has been arrested at an early stage of development (Fujita et al. 1966; Schmechel and Rakic 1979b) or whether mature glia undergo divisions? An insight into this problem was gained by the combined application of immunocytochemistry and autoradiography. Our results suggested that the radial glia, being capable of repeated divisions while having processes anchored to the pial surface, meet the criteria of a common precursor that may give rise to a number of cell types found in proliferation at nongerminal sites within the period studied. This implies that in the forebrain, similar to that seen with routine histological technique in the cerebellum, glial cells may multiply even in fairly advanced stages of differentiation. This raises the question of whether in the case of the glia the classical proliferation-migration-differentiation sequence of nervous tissue cell maturation makes sense at all. If a cell can divide in any position and age, when can it be regarded as fully mature or differentiated? Our answer to these questions is that a glial cell is mature when it occupies its final position and develops its arbor of processes. Hence the basic difference in the course of neurogenesis and gliogenesis is that, while in developing neurons the phases of proliferation, migration, and differentiation are strictly consequent, glial cells may proliferate even after their differentiation into one or another cell type has been accomplished. Another interesting feature of glial proliferation is that it does not necessarily require any kind of dedifferentiation. While in most cell types renewed proliferative activity is accompanied by a marked dedifferentiation, for example, the loss of processes and the rounding of the cell body in the case of neurons, proliferative glial cells were seen to retain processes and to possess almost all the structural features thought to characterize mature astrocytes (Glees and Meller 1968; Peters et al. 1970).

Therefore, there seems to be no need to suppose the presence of dormant, undifferentiated stem cells to explain the retained postnatal proliferation capacity of the glia as the radial glia, and its derivatives clearly show such potentialities.

7.4 Derivatives of and Mechanism of Derivation from the Radial Glia

7.4.1 Astrocytes

What kind of glia may originate from the radial glia? The population suggested already by Ramón y Cajal (1890, 1911) and more recently by Akers (1977) and Schmechel and Rakic (1979a) to be derived from this embryonic glia was the astrocytes. However, no suggestion has been made concerning the mechanism of astrocyte formation from the radial glia. From out studies it appears that the radial glia are withdrawn by repeated divisions. The predominant orientation of mitotic spindles observed permits us to predict that after these mitoses one daughter cell remains in possession of the outwardly directed processes contacting the pial surface, while the other is left behind in the growing neuropil. These latter may differentiate into astrocytes, migrate away, or remain in their original form with a centrally directed process. The cell still contacting the pial surface is virtually shorter than in previous stages as the brain has increased in size in the meanwhile. A subsequent mitosis may divide again the remnant of the radial glia in the same manner as described above. Divisions may either proceed until the cell contacting the pial surface with a long process is cut to a subpial astrocyte or stop earlier yielding a glial cell with processes of varying lengths.

It should be emphasized that this mechanism may not be the only way of astrocyte formation. As our concern was the radial glia rather than astrocyte formation in general, the consideration of other possible sources of astrocytes is beyond our present scope.

7.4.2 Cerebellar Bergmann Glia

The mechanism proposed above is clearly demonstrable in the cerebellum (Bascó et al. 1977). The development of the Bergmann glia exemplifies the case when the retrieval by divisions of the radial glia is incomplete and proliferation is arrested at a stage when radial processes (the Bergmann fibers) are not yet disintegrated. Thus the Bergmann glia may be regarded as a permanent derivative of the rhombencephalic radial glia.

At other sites of the cerebral hemispheres only subpial astrocytes remain, with relatively short processes contacting the surface. They are particularly well seen along the interhemispheric fissure.

7.4.3 Diencephalic Tanycytes and Retinal Müller Cells

In the diencephalon the radial glia undergo only minor modifications as compared with other forebrain areas. This fits in with the finding that between P1 and P14 neither mitotic figures nor [³H]thymidine-incorporating cells were observed in the radial glia of this region. Although Wenger et al. (1966) have reported on mitoses in the adult rat diencephalon, their relationship with glial elements could not be verified. Tanycytes of the adult brain are thus considered to be the closest derivatives of the radial glia and to have much in common with the original embryonic asymmetrically bipolar radial glial cells.

On theoretical grounds the Müller cells of the retina were also classified into this group (Friede 1965; Varon and Somjen 1979). The retina bulges out from the presumptive diencephalon, and Müller cells whose glial nature is unequivocally agreed (Müller 1851; Ramón y Cajal 1904, 1909; Polyak 1941, 1956) span the retina (i.e., the modified diencephalic wall) exactly as the embryonic radial glia at other parts of the developing central nervous system. In addition to these morphological and developmental similarities Ikeda et al. (1980) reported on the GFAP immunostaining of Müller cells in the chicken retina, a powerful argument for regarding Müller cells as part of the diencephalic radial glia.

7.5 Common Properties of Radial Glial Derivatives

We have thus pointed out a possible common precursor for different neuroglial cell populations. But where should these populations be placed within the generally accepted classification of glia?

A straightforward answer for this question was provided by the demonstration of a common immunospecificity shared by all cell types studied. The GFAP immunostaining carried out in this work in the forebrain and by Bignami and Dahl (1974a, b) in the cerebellum indicates that they belong to the class of astroglia.

There are, however, several other properties which are invariably common in these radial glia-derived GFAP-immunopositive populations.

7.5.1 Capability of Transporting Material Between Various Brain Fluid Spaces

Brain fluid spaces are separated from each other by functional barriers constituted of nonneuronal elements such as endothelial cells, basement laminae, and glial limitans membranes. Accordingly, glial cells appear to have access to these fluid spaces. Radial glial cells in particular, by having processes ending with end feet at pial and/or ventricular surfaces, seem to be capable of connecting two distant fluid spaces. Transport of material along the radial glia fibers has been verified in various parts of the brain, especially in the diencephalon (Rodriguez 1972; Léránth and Schiebler 1974; Scott et al. 1974; Wagner and Pilgrim 1974; Nakai and Naito 1975; Hajós et al. 1982) and now in the forebrain and cerebellum.

The functional significance of this mechanism is obvious in the diencephalon, where the system of tanycytes may provide the structural basis of neurohumoral transport from cerebrospinal fluid to brain extracellular space or vice versa [see Palkovits (1978) for references]. A similar communication via tanycytes may exist between blood vessels and the ventricle (Rodriguez 1972). Less clear is the significance of such a transport mechanism in the case of the telencephalic and rhombencephalic radial glia. Stensaas and Gilson (1972) and Hinds and Hinds (1972) have proposed an information transfer by transported macromolecules from the brain surface to the sites of cell proliferation. Modulation of glial behavior by the transport of surface molecules has also been suggested (Gaze and Watson 1968). These assumptions await further experimental support.

7.5.2 Guidance of Neuronal Migration

Rakic has put forward his hypothesis (Rakic 1971a, b, 1972) that migrating neurons are guided by glial fibers. The general validity of this hypothesis has been argued ever since. The claim of Rakic was based on a set of observations of migrating neurons. Neurons in this phase acquire an elongated shape vertical to the surface and move across the intricate network of neuropil in a strictly straight radial direction. It seemed reasonable to assume that such a straight path of migration is impossible unless preformed routes for cell movements are available. From this assumption it was only one step further to propose that the radial glia with its fibers neatly perpendicular to the brain surface provide these preformed routes for neuronal migration. Accordingly, migratory neurons would attach to glial fibers, which would then guide them right to their final place. Rakic has described this mechanism for the monkey neocortex and cerebellar Bergmann fibers and extrapolated its operation to the whole brain (Rakic 1971 a, b, 1981).

Attractive as this hypothesis of contact guidance is, it has immediately been challenged on the developmental ground that the Bergmann glia are generated later than the peak of neuronal migration from the external layer of the cerebellum occurs. On the basis of thymidine autoradiography the Bergmann glia was claimed to be generated on P8–10 (Fujita et al. 1966; Das et al. 1974), while masses of granule cells are known to descend from the external granular layer from P4–5 onward.

However, the basis for this challenge turned out to be wrong. Using different techniques, Del Cerro and Swarz (1976), Swarz and Oster-Granite (1978), and Choi and Lapham (1978) have clearly demonstrated the presence of radial fibers in both the cerebrum and cerebellum at embryonic stages prior to the onset of massive neuronal migration. The work of Bascó et al. (1977) provided evidence for the proliferation of glial cells in the cerebellum that have long processes extending to the pial surface, and a mechanism of formation of the Bergmann glia has been proposed. It is, in fact, only a matter of terminology what is called the Bergmann glia: is it all the cerebellar glial cells that have processes extended toward the surface or only those whose perikarya are situated in the layer of Purkinje cells? Regardless of the question of maturity it appears fairly obvious that as far as guidance of neuronal migration is concerned all cells with radial process(es) may be considered as equivalent to the Bergmann glia.

With GFAP immunostaining we were able to demonstrate a close association of migrating neuroblasts with radial fibers in the cerebellar cortex, cerebral cortex, and hippocampal formation. As to the diencephalon the situation could not be assessed because tanycytes at an embryonic age when neuronal migration occurred did not stain with the GFAP reaction. So investigation of the participation of tanycytes in neuronal migration is still pending.

The question of guidance of neuronal migration by glial fibers is of particular interest in the cerebral cortex, the structural organization of which (Mountcastle 1957, 1978; Von Bonin and Mehler 1971; Hubel and Wiesel 1962, 1977; Szentágothai 1975) may well justify a development along radial tracks. However, our systematic analysis of the GFAP-immunopositive radial glia in the cerebral cortex showed an uneven distribution of GFAP-positive fibers. For example, no GFAP immunoreactivity could be demonstrated in the occipital cortex. Moreover, in many areas the cortical plate was present prior to the appearance of GFAP immunopositivity (Woodhams et al. 1982). Of course, one must be extremely cautious not to overestimate these findings because they apply only for the appearance of a specific protein; the primary presence of radial glial fibers cannot be ruled out. That methodical factors and species differences may also be involved is suggested by the finding of GFAP-immunopositive radial fibers in the occipital cortex of the rat by Wolff (personal communication). Unfortunately, recent findings relevant to the issue of contact guidance by the radial glia have not brought the problem nearer to its solution. For example,

Hatten and Liem (1981) claimed that even in culture the astroglia provides a template for positioning cerebellar neurons, while this could not be confirmed in the studies of Balázs (personal communication).

Bearing all this in mind the developmental course of the radial fiber system still suggests that the mechanism of contact guidance of neuronal migration may be operational in the acquisition of a later generation of neurons by the cortical plate. This is consistent with the view (cf. Jacobson 1978) that early formed, temporarily bipolar neurons may migrate by means of their own processes attached to the outer surface. Later-formed neurons do not have such processes; therefore, their migration through the labyrinth of the gradually increasing neuropil may require guidance.

7.5.3 Morphological Similarities

All cell types supposed to be derived from the radial glia have common histological features. The nuclear structure is invariably of the astrocyte type in all radial glial derivatives: round or oval with light, sparse evenly distributed chromatin. Under the light microscope these nuclei appeared to be pale as compared with the dark nuclear staining of small neurons and the oligodendroglia but not as light as the nuclei of some large neurons. A characteristic feature of the nucleus was its well-outlined nuclear envelope. Under the electron microscope this appeared to be due to a fine apposition of chromatin at the inner aspect of the inner nuclear membrane. The cytoplasm proved to be also of the "watery" astrocytic type in all radial glial derivatives studied. Organelles were few in number. They were found scarcely distributed in the cytoplasm. Mainly mitochondria and some rough-surfaced endoplasmic cisterns occurred.

In contrast to the observation of Chu Wong et al. (1981) in anuran larvae, during the perinatal period of the mouse glial fibrils or filaments typical for mature astrocytes were scarcely seen either in the cell body or in the processes of the radial glia and derived cells of the brain. This would not be surprising had it not been coincident with the presence of a marked GFAP immunoreactivity in the same cell type. GFAP immunoreactivity, as reflected by its name, was thought to be associated with the glial fibrils (Bignami and Dahl 1973, 1974a, b, 1975) known to appear in astroglial cells at a certain stage of maturation. It seems that GFAP-immunoreactivity is detectable earlier than fibrils develop. This may be due to several reasons. The astrocyte-specific protein may be synthesized within the cell prior to the formation of fibrils with which it may associate later. Nor can it be excluded that astrocyte-specific protein is not only associated with the fibrils but also distributed throughout the whole cytoplasm. Of course if this were the case, the term "fibrillary" in the full descriptive name of this protein would need revision.

These problems could most adequately be approached by the electron microscopic demonstration of GFAP immunoreactivity in the course of development. Such studies have recently been carried out by Levitt et al. (1981) in the monkey.

Another aspect of cellular morphology, the ramification pattern, showed variations from type to type in the radial glial derivative populations.

The original asymmetrically bipolar form of the embryonic radial glia with a short central, and a long peripheral, process is best reflected by the shape and process arrangement of diencephalic tanycytes. The other diencephalic radial glia-derived cell type, the Müller cells of the retina, differs from the original form by a less asymmetrical shape: the two processes are almost equal in length.

In the cerebellum the radial glia undergo a structural rearrangement: by means of repeated mitoses in a plane parallel to the surface the outer daughter cell retains the radial process whereas the inner daughter cell loses contact with the pial surface and develops into a stellate astrocyte of the granular layer and white matter. In the meanwhile the outer daughter cell — the Bergmann glia cell — differentiates further and acquires its mature form with a number of radial processes (the Bergmann fibers) and a few short processes directed toward the granular layer.

In the telencephalon a type of radial glia reminiscent of the cerebellar Bergmann glia was encountered in the dentate gyrus. Their perikarya were located in one single row beneath the granular layer of the gyrus. The long radial process could unequivocally be verified; however, the existence of centrally directed processes or extensions is still obscure.

Little, if anything, can be said about the shape and process pattern of the persistent radial glia of the ventrolateral corner of the lateral ventricle. The presence of this type of glia was only indicated by GFAP immunocytochemistry and HRP uptake, but much further work and technical sophistication is required to elucidate the exact shape and ramification pattern of these cells.

An important aspect of morphology of radial glial derivatives is the pattern of terminal arborization. The embryonic radial glia have a characteristic brush- or fork-like terminal arbor, each branch ending in a club-shaped endfoot under the pia. Except for the telencephalic radial glia, which shows a roughly similar terminal arbor, no such typical terminal ramification is seen in the radial glia derivative cell types. As a rule, long processes end in a single endfoot without ramification. This single endfoot is often bulky, as in the case of retinal Müller cells whose endfeet limiting the layer of optic fibers may even reach the size of the cell body.

8 Concept of the Surface-Contact Glia

In the foregoing we have considered the developmental, immunocytochemical, and morphological similarities that support the view that the cell types studied belong to the same class of glia. This class appears to be astrocytic in nature but it obviously comprises more cell types than conventional astrocytes. An attempt to classify these cell types on a common basis has already been made by Friede (1965), but lacking wider developmental and experimental support it has remained at the level of a working hypothesis.

The present study is thought to provide the necessary evidence, both functional and structural, to substantiate the view of a new, more comprehensive classification of certain radial glia-related populations. The question is which of the common properties may be suitable to define the entire group? Developmental aspects such as the course and timing of proliferation reflect only the embryonic or early postnatal situations, as after these periods glial proliferation in the fully mature, intact brain is sporadic. Immunocytochemistry has proved to be extremely useful in pointing out chemical similarities, but as discussed in the previous chapter (7.5.3) the lack of immunoreactivity does not necessarily mean the lack of radial glia-related elements.

All this implies that the feature we can use to specify the glial population shown to be similar developmentally and immunocytochemically, should be sought among the morphological landmarks, the more so since these landmarks also characterize the population in the adult brain.

There appears to be one feature fundamentally common to the morphology of the radial glia-derived cells: *all of them contact a brain surface by at least one of their processes*. The surface may be either pial or ventricular but in either case access to a cerebrospinal fluid space (ventricles or subarachnoid space) is ensured. Thus we term all radial glia-derived populations shown in this study to share a number of common developmental, immunocytochemical, morphological, and functional properties, as the *surface-contact glia*.

On the basis of morphological, autoradiographic, immunocytochemical, and functional evidence a new comprehensive classification of the astroglia is proposed. Accordingly, the astroglia comprise two main groups of cells derived from a common precursor, the embryonic radial glia:

1. Mature astrocytes of the deep neuropil having the functions ascribed generally to the astroglia, i.e., isolation of neuronal elements, space-filling and cementing, participation in neuronal metabolism, phagocytotic activity, etc.

2. The surface-contact glia, consisting of two subpopulations:

a) Transforming type (most of the embryonic radial glia of the telencephalon)

b) Permanent types (subpial astrocytes, the radial glia of the ventricular corner, the hippocampal radial glia, diencephalic tanycytes, Müller cells of the retina, and the cerebellar Bergmann glia)

The surface-contact glia may have the functions of establishing communication and material transport between various brain fluid spaces, guiding neuronal migration at certain sites and developmental stages, and transferring information from the brain surface to places of cell proliferation. Its embryonic form may give rise to astrocytes.

This interpretation and classification of radial glia-derived cell populations broadens the term of astroglia by regarding, on the basis of evidence presented, a number of glial or hitherto unclassified cell types as belonging to the class of astroglia. It also has implications as to the general course of cell acquisition in the developing central nervous system:

1. There is marked postnatal mitotic activity at various nongerminal sites of the brain which in the first two postnatal weeks is clearly related to the withdrawal by repeated divisions of the widespread radial glia system.

2. This postnatal proliferative activity has a well-defined regional program with regular onset, peak, and cessation times. This process follows with a certain lag the time course of regional withdrawal of the radial glia, although it may be incomplete in some areas and various cell types that resemble more or less the embryonic radial glia are left behind.

3. The general rule that neurons are produced before the glial cells of their region does not apply for the radial glia, which is one of the ontogenetically oldest cell types of neuroepithelial origin.

4. The contact guidance of migrating neuroblasts by glial fibers is a feasible means of development of geometrically organized brain areas such as the central and cerebellar cortices. However, our results suggest that this mechanism is operational in the migration of the later-generated neurons, which in contrast to the early ones do not have their own processes and may need guidance to migrate through the intricately developed neuropil.

9 Current Approaches to the Glia and Some Perspectives

The development of knowledge on the glia, as outlined briefly in Sect. 1.1, has now reached a stage where it has become evident that no one method used on its own can extend our understanding of glial structure and function. On the other hand, powerful new techniques such as autoradiography, immuno-cytochemistry at both light and electron microscopic levels, and tracer adminis-trations have added substantially to the classical histological methods, which have thus reached a high level of sophistication. The combination of this new type of histology with biochemical and functional investigations is thought to determine glial research of the near future.

In this monograph we have tried to approach our subject by adopting a number of up-to-date techniques, all of them fundamentally histological in na-ture, i.e., it was finally the microscope — with either light or electron optics — under which results were evaluated. Under these circumstances it is clear that the conclusions drawn are mostly of a histological or developmental nature. Obvious as the need is to combine these techniques with biochemical and physio-logical methods, its realization presents great difficulties. Any major break-through in this territory can be expected only by the clarification of some questions inherent to the biochemistry and physiology of the glia.

The first among these is what kind of preparation should be used for the biochemical and/or physiological analysis of the glia? In the literature of recent years two major trends can be observed. One is the use of glial cells derived from glial tumors (reviewed by Wollemann 1974; Giacobini et al. 1980); the other is the bulk separation of glial cells from brain tissue (reviewed by Podluslo and Norton 1972; Rose 1969; Hamberger and Sellström 1975). Although both preparations have proved to be most useful in elucidating some basic phenom-ena, neither is without its own difficulties.

With respect to tumor cells it is always a problem how much of the original glial phenotype is expressed in them. One may argue that, for example, endocrine tumor cells produce high amounts of the same type of hormone as their normal counterparts and thus they can be regarded as functionally comparable to them, but in the case of the glia the capability of proliferation is — as shown also in this study — a fundamental property the rate of which is rather characteristic for the subtypes of the glia. Unfortunately, it is this cell function which is fundamentally altered in tumor cells. In spite of this, the tissue culture of glial tumor cells is a widely used preparation and it will certainly remain so until other types of preparations offer a better alternative.

For some time, such a better alternative has been claimed to have been found in the bulk separation of glial cells. The first steps in separating glial cells involved the disruption of the nervous tissue so that glial cell bodies could be brought into a solution. The disruption, using either the enzymic digestion of intercellular connections or simply mechanical sieving, resulted in the amputation of glial processes. Once cell bodies were isolated and suspended in a medium their separation by various sedimentation procedures became merely a technical problem. This type of approach yielded unequivocal results when applied to central nervous system areas containing only glial cells such as the corpus callosum or other large commissural bundles.

Our present concern, however, is the radial glia and derivative cell types, all of them located in areas containing masses of neurons. Curiously enough biochemists who have carried out pioneering work on the bulk separation of the glia did not and do not seem to be worried about possible difficulties in separating glial and neuronal populations from a mixed suspension. This confidence emerges from the postulate that neurons are large, while glial cells are small, a view that the neurohistologist cannot share. It is true that large neurons (cortical and hippocampal pyramidal cells, cerebellar Purkinje cells, motor nerve cells of cranial nerve nuclei, etc.) are larger than glial cells, especially those of the oligodendroglia. However, the population of the so-called microneurons is far too large to be neglected. There are, for example, estimates (Elliott 1969) according to which the cerebellar granule cells − typical microneurons − outnumber all nerve cells in the brain. Of course many of these quantitative estimates of central nervous system cell numbers have to be regarded with caution, the more so since some figures published are almost immediately challenged, while others prove to be good examples of the "scientific folklore:" everybody quotes them and nobody knows where they come from! However, on balance one may still conclude that to regard the population of small cells as purely glia is a rough oversimplification.

A more scrupulous attempt has been made by Balázs et al. (1979) to isolate glial and neuronal populations from the immature cerebellum. While they were able to identify morphologically the population of large neurons, the population of small cells was further resolved by unit-gravity sedimentation into subpopulations each having been subjected to careful morphological, biochemical, and immunocytochemical analysis. On this basis the population of small cells proved to be constituted of granule cells (from both the external and internal granular layers) and astrocytes. These latter could be separated (Cohen et al. 1978) and their viability was tested by metabolic studies and by their successful culturing. An additional advantage of this procedure over methods using white matter was that it provided equally viable neuronal populations that could be studied comparatively with the astrocyte population. Moreover, the successful culturing of the isolated populations resulted in the regrowth of amputated processes. Thus this preparation seems to be a good alternative to using tumor cells either for biochemical analysis or for studies in tissue culture. The method of Balázs et al. (1979) has, however, the limitation that the dispersion of tissue yielding viable cells can be carried out only in immature animals (in rats not older than 14 days). Taking into consideration that the main events of neuronal-glial interaction during development, in particular those related to the radial glia, take place predominantly before this period (reviewed by Rakic 1981), the experimental approach of Balázs et al. (1979) may be applicable to areas where the radial glia appears to play an important role in brain development.

Although from what has been discussed in this chapter the reader may well feel that the authors' preference is a sort of combination of the full arsenal of modern histology with the biochemical and functional analysis of bulk-separated glial cells as a promising future approach to a more profound understanding of the radial glia, it should not be overlooked that glial research going on its own way has also produced other remarkable interdisciplinary combinations that have greatly contributed to the advance of knowledge and still continue to do so. For example, the autoradiographic study of isolated ganglia (Schon and Kelly 1974; Iversen et al. 1975) provided valuable information on transport processes of glial cells. Results could be extrapolated to the central nervous system glia serving as a guideline to finding specific markers for glial cells. Recently Garthwaite et al. (1980) reported on the $1^1/_2$- to 2-h incubation of specially cut cerebellar tissue slices that have preserved their full structural and ultrastructural integrity. This type of preparation is most promising even for the direct functional-morphological observations of various cell populations including the glia.

Finally, when trying to envisage future trends it is not difficult to predict that immunological methods will play a central role in glial histology as they do in a number of other fields. Immunocytochemistry has revitalized light microscopic cytochemical research and its electron microscopic application has opened new perspectives. As far as the glia are concerned immunocytochemistry is still far from its climax, as to date the only widely used glial-specific protein is GFAP, but many more can be expected to be introduced in the near future in routine glial histology. Promising results are obtained, for example, with the fibroblast filament-specific protein, vimentin (Dahl et al. 1981), which appears to be present in the radial glia earlier than GFAP (Woodhams 1983, personal communication). In possession of antisera against oligodendroglial-specific proteins, or antisera against proteins specific for various subtypes of the radial glial line, a more comprehensive analysis of the role of the glia in central nervous system development can hopefully be performed.

10 Summary

The present study was focused on a cell population of the immature brain present from the earliest stages of neural tube formation. This population was observed by last century neurohistologists as asymmetrically bipolar cells having a short process contacting the ventricular lumen and a long process directed toward the outer surface and terminating in end feet under the pia. The terminology of this cell type reflects a great deal of ambiguity concerning its role and function. The most recent term "radial glia" has been introduced to denote one of its characteristic morphological features: long straight processes spanning the wall of the neural tube in a direction neatly perpendicular to the outer surface. Owing to the virtual disappearance of this cell type from the mature nervous system, relatively less attention has been paid to it with regard to the mechanism of withdrawal and possible lineages.

Data published in the past decade suggest that the investigation of this embryonic glia may provide a clue to several poorly understood aspects of gliogenesis and related events of brain development such as:
1. The significance of postnatal mitotic activity at the nongerminal sites of the brain
2. The regional distribution and time course of glial proliferation in the brain
3. The temporal relationship of gliogenesis and neurogenesis
4. The participation of the glia in the structural organization of geometrically arranged brain areas
5. How to define the astroglia?

These questions were investigated in mice from early embryonic (E) age until various postnatal (P) periods, using a variety of histological techniques including light microscopy of toluidine-blue-stained epoxy resin sections, routine electron microscopy, light and electron microscopic autoradiography of [^{3}H]thymidine incorporation, light microscopic immunocytochemistry of the glial fibrillary acidic protein (GFAP), Golgi silver impregnation, and the retrograde transport of the tracer horseradish peroxidase (HRP).

An atlas of postnatal cell proliferation at nongerminal sites of the brain was prepared by mapping the occurrence of mitotic figures from P1 to P12 in frontal serial sections of the resin-embedded brain. The subventricular zones, the external granular layer of the developing cerebellum, the granular layer of the dentate gyrus, and the core of the olfactory bulb were regarded as the primary and secondary germinal sites and were therefore not evaluated.

In the immature cerebellum mitotic figures were observed under the layer of Purkinje cells (the presumptive internal granular layer), reaching the highest number between P7 and P12. Mitotic cells showed a clear-cut connection with the radial fiber system of the primitive molecular layer. Fibers originated from the apical pole of the mitotic cells, and the pattern of their arborization corresponded to that characteristic for the Golgi picture of Bergmann glia cells.

In most cases mitoses took place in a plane parallel to the surface. Mitotic cells were thus identified as the Bergmann glia.

In the forebrain mitotic cells were seen in a number of areas. Mitotic activity appeared to be the highest in the hippocampus between P0 and P7, in the corpus callosum between P0 and P3, in the caudate-putamen on P3, in the frontal cortex on P3, and in the sensorimotor cortex between P5 and P7. Moderate or low mitotic activity was observed in the pyriform cortex and the limbic areas between P0 and P12, and in the occipital cortex between P5 and P7. Mitotic activity ceased by P12 at all nongerminal sites studied.

The distribution and time course of the mitotic activity and the morphological appearance of mitotic cells suggest that this wave of cell proliferation corresponds to the repeated divisions of the astroglia. Findings in the cerebellum indicate that this cell multiplication may be related to divisions of the radial glial cells.

To corroborate this assumption immature mice were *pulse-labeled with [³H]thymidine* and the proliferative activity was studied by light and electron microscopic autoradiography. The occurrence of labeled cells showed a roughly similar pattern to that of mitotic figures. Under the electron microscope two types of labeled cells could be distinguished. On the basis of morphological criteria they were identified as astroglial precursors.

The relationship between postnatal, nongerminative mitotic activity attributed to gliogenesis and the radial glia was investigated using *silver impregnation* and *GFAP immunocytochemistry*. The appearance and disappearance of GFAP immunoreactive fibers showed a regular course throughout the brain. In the cerebral cortex immunopositive fibers were seen first in the frontal cingulate cortex at E17 and they were following a rostrocaudal and dorsoventral gradient of appearance and disappearance within the cortex so that after P20 the area around the stria terminalis (ventrolateral angle of the lateral ventricle) was the only place where they could be demonstrated.

A consistent finding was the lack of correlation between the pattern of cortical plate formation and the appearance of GFAP-positive radial fibers. It is of interest that where radial fibers were traversing areas containing migratory neurons they appeared to be closely associated with these migratory cells.

In the hippocampus GFAP-positive radial fibers were encountered from E16 onward. They persisted until P9. Radial glial fibers in the dentate gyrus remained strongly stained and persisted even after P30. Glial perikarya in the dentate gyrus were arranged in a single row under the granular layer.

In the diencephalon, tanycytes showed a strong GFAP immunoreactivity from E17. Their location varied with age: first they were seen around the floor of the third ventricle; later they gradually surrounded the whole ventricle.

Findings indicate that the radial glia of the developing cortex contain an astrocyte-specific protein and can thus be regarded as an astroglial cell type. This immunospecificity is also shared by the radial glia of the hippocampus and dentate gyrus, the cerebellar Bergmann glia, and diencephalic tanycytes. This together with developmental and morphological observations provides circumstancial evidence that the early postnatal mitotic activity of nongerminal sites of the brain is related to the divisions of the radial glia which, in turn, appears to be the precursor of the astroglia and a number of other cell types.

The simultaneous demonstration of [³H]thymidine incorporation by GFAP-

immunopositive cells provided direct evidence to support our claim that the radial glia and their derivatives retain their proliferative capacity over long postnatal periods and are capable of dividing even after having developed long processes anchored to a brain surface.

Studies with the *retrograde transport of HRP* have clearly revealed that the transport of materials along glial processes is a function common in all radial glia-derived cell types.

Taking into account the developmental, morphological, immunocytochemical, and functional parameters studied the following conclusions are drawn:

1. The marked postnatal mitotic activity at various nongerminal sites of the mouse brain involves the disintegration of the embryonic radial glia by repeated mitoses.

2. This postnatal proliferation has a well-defined regional program with regular onset, peak, and cessation periods and follows roughly the regional withdrawal of the radial glia. There remain, however, places where this withdrawal is incomplete.

3. The general rule that neurons are produced before the glial cells of their regions does not apply to the radial glia, which is one of the ontogenetically oldest cell types of the neuroepithelium.

4. The guidance by glial fibers of migrating neuroblasts is a feasible means of development of geometrically organized brain areas. However, the discrepancy between cortical plate formation and the appearance of the radial glia suggest that this mechanism is operational in the migration of the later-generated neurons, which in contrast to early neurons do not have their own processes and have to migrate through an intricately developed neuropil.

As a main concept emerging from our studies complemented by some data available in the literature, a new classification of the astroglia population is proposed based on the demonstrated lineage from the radial glia. This new, broader class of astroglia is constituted by two main groups of cells derived from a common precursor, the embryonic radial glia. One is the astrocytes of the deep neuropil, while the other is specified by a feature common in all of its members throughout their whole life: the possession of at least one process contacting a brain surface. This *surface-contact glia*, astrocytic in nature, comprises a transforming type — most of the radial glia of the telencephalon — and permanent types such as the subpial astrocytes, the radial glia of the ventricular angle, the dentate gyrus radial glia, tanycytes, Müller cells of the retina, and the cerebellar Bergmann glia.

The surface-contact glia may have the functions of establishing communication between various brain fluid spaces, guiding neuronal migration at certain sites and developmental stages, and transferring information from the brain surface to places of cell proliferation. Its embryonic form may give rise to astrocytes.

Acknowledgments

Some of the investigations contained in this study were made in collaboration with Drs. Robert Balázs and Peter Woodhams from the MRC Developmental Neurobiology Unit, London; Kyrill Reznikov from the People's Friendship University, Moscow; and Zoltán Fülöp and András Csillag from the Semmelweis University, Budapest, whose contribution is greatly appreciated.

11 References

Akers RM (1977) Radial fibers and astrocyte development in the rat cerebral cortex. Anat Rec 187:520

Allen F (1912) The cessation of mitosis in the central nervous system of the rat. J Comp Neurol 22:547–568

Altman J (1962) Autoradiographic study of degenerative and regenerative proliferation of neuroglia cells with tritiated thymidine. Exp Neurol 5:302–318

Altman J (1966a) Proliferation and migration of undifferentiated precursor cells in the rat during postnatal gliogenesis. Exp Neurol 16:263–278

Altman J (1966b) Autoradiographic and histological studies of postnatal neurogenesis. II. A longitudinal investigation of the kinetics, migration and transformation of cells incorporating tritiated thymidine in infant rats, with special reference to postnatal neurogenesis in some brain regions. J Comp Neurol 128:431–473

Altman J (1968) DNA metabolism and cell proliferation. In: Lajtha A (ed) Handbook of neurochemistry. Plenum, New York, Vol 2, pp 137–182

Altman J (1969) Autoradiographic and histological studies of postnatal neurogenesis. III. Dating the time of production and onset of differentiation of cerebellar microneurones in rats. J Comp Neurol 136:269–294

Altman J (1975) Postnatal development of the cerebellar cortex in the rat. IV. Spatial organization of bipolar cells, parallel fibres and glial palisades. J Comp Neurol 163:427–448

Altman J, Das GD (1964) Autoradiographic examination of the effects of enriched environment on glial multiplication in the adult rat brain. Nature 204:1161–1165

Altman J, Das GD (1965) Autoradiographic and histological evidence of postnatal hippocampal neurogenesis in rats. J Comp Neurol 124:319–336

Altman J, Das GD (1967) Postnatal neurogenesis in the guinea-pig. Nature 214:1098–1101

Angevine JB (1965) Time of neuron origin in the hippocampal region: an autoradiographic study in the mouse. Exp Neurol [Suppl] 2:1–70

Angevine JB, Sidman RL (1962) Autoradiographic study of histogenesis in the cerebral cortex of the mouse. Anat Rec 142:210

Antanitus DW, Choi BH, Lapham LW (1976) The demonstration of glial fibrillary acidic protein in the cerebrum of the human fetus by indirect immunofluorescence. Brain Res 103:613–616

Ariens Kappers CU, Carl Huber G, Caroline Crosby E (1960) The comparative anatomy of the nervous system of vertebrates including man. vol III. Hafner, New York (Originally published in 2 volumes in 1936)

Athias M (1897) Recherches sur l'histogenese de l'ecorse du cervelet. J Anat (Paris) 33:372–404

Balázs R, Cohen J, Woodhams PL, Patel NJ, Garthwaite J (1979) Isolation and characterization of cell types from the developing cerebellum. In: Meisani E, Brazier MAB (eds) Neural growth and differentiation. Raven, New York, pp 113–132

Bartlett PF, Nobel MD, Pruss RM, Raff MC, Rattray S, Williams CA (1981) Rat neural antigen-2 (RAN-2); a cell surface antigen on astrocytes, ependymal cells, Müller-cells and lepto-meninges defined by monoclonal antibody. Brain Res 204:339–351

Bascó E (1981) Regional distribution and time course of mitotic activity of astroglia in the immature mouse forebrain. Acta Morph Acad Sci Hung 29:203–211

Bascó E, Hajós F, Fülöp Z (1977) Proliferation of Bergmann-glia in the developing rat cerebellum. Anat Embryol (Berl) 151:219–222

Bascó E, Woodhams PL, Hajós F, Balázs R (1981) Immunocytochemical demonstration of glial fibrillary acidic protein in mouse tanycytes. Anat Embryol (Berl) 162:217–222

Bayer S, Altman J (1974) Hippocampal development in the rat: cytogenesis and morphogenesis examined with autoradiography and low-level X-irradiation. J Comp Neurol 158:55–80

Bellongi G, Stefani A (1889) Contribution a l'histogenese de l'ecorce cerebellaire. Arch Ital Biol 11:21–25

Berry M (1974) Development of the cerebral neocortex in the rat. In: Gottlieb D (ed) Studies on the development and behaviour and the nervous system. Vol 2, Aspects of neurogenesis. Academic, New York

Bignami A, Dahl D (1973) Differentiation of astrocytes in the cerebellar cortex and the pyramidal tracts of the newborn rat: an immunofluorescence study with antibodies to a protein specific to astrocytes. Brain Res 49:393–402

Bignami A, Dahl D (1974a) Astrocyte-specific protein and radial glia in the cerebral cortex of newborn rat. Nature 252:55–56

Bignami A, Dahl D (1974b) Astrocyte-specific protein and neuroglial differentiation: an immunofluorescence study with antibodies to the glial fibrillary acidic protein. J Comp Neurol 153:27–38

Bignami A, Dahl D (1975) Astroglial protein in the developing spinal cord of the chick embryo. Dev Biol 44:204–209

Boulder Committee (1970) Embryonic vertebrate central nervous system, revised terminology. Anat Rec 166:257–262

Brückner G, Mares V, Biesold D (1976) Neurogenesis in the visual system of the rat. An antoradiographic investigation. J Comp Neurol 166:245–256

Bunge RP (1968) Glial cells and the central myelin sheath. Physiol Rev 48:197–251

Bunge RP (1970) Structure and function of neuroglia: some recent observations. In: Schmitt FO (ed) The neurosciences, second study program, Rockefeller University Press, New York, pp 782–797

Caley DW, Maxwell S (1968) An electron microscopic study of the neuroglia during postnatal development of the rat cerebrum. J Comp Neurol 133:45–70

Cammermeyer J (1963) Differential response of two neurone types to facial nerve transection in yound and old rabbits. J Neuropathol Exp Neurol 12:203–211

Cavanagh JB (1970) The proliferation of astrocytes around a needle wound in the rat brain. J Anat 106:471–487

Caviness VS, Rakic P (1978) Mechanisms of cortical development: a view from mutations in mice. Ann Rev Neurosc 1:297–326

Caviness VS, Pinto-Lord MC, Everard P (1981) The development of laminated pattern in the mammalian neocortex. In: Conelly TG, Brinkley LI, Carlson BM (eds) Morphogenesis and pattern formation. Raven, New York, pp 103–126

Chiu FC, Norton WT, Fields KL (1981) The cytoskeleton of primary astrocytes in culture contain actin, glial fibrillary acidic protein and the fibroblast-type filament protein, vimentin. J Neurochem 37:147–155

Choi B, Lapham L (1978) Radial glia in the human fetal cerebrum: a combined Golgi, immunofluorescent and electron microscopic study. Brain Res 148:295–311

Chow KL, Dewson JH (1966) Numerical estimates of neurones and glia in lateral geniculate body during retrograde degeneration. J Comp Neurol 128:63–74

Chu Wong JW, Oppenheim RW, Farel P (1981) Ultrastructure of migrating spinal motoneurons in anuran larvae. Brain Res 213:307–318

Cohen J, Balázs R, Hajós F, Currie DN, Dutton GR (1978) Separation of cell types from the developing cerebellum. Brain Res 148:313–331

Dahl D, Rueger DC, Bignami A, Weber K, Osborn M (1981) Vimentin, the 57000 dalton protein of fibroblast filaments, is a major cyto-skeletal component in immature glia. Eur J Cell Biol 24:191–196

Dalton MM, Hommes OR, Leblond CP (1968) Correlation of glial proliferation with age in the mouse brain. J Comp Neurol 134:397–400

Das GD (1976) Differentiation of Bergmann glia cells in the cerebellum: a Golgi study. Brain Res 110:199–213

Das GD, Lammert GL, McAllister JP (1974) Contact guidance and migratory cells in the developing cerebellum. Brain Res 69:13–29

Davidoff M (1973) Über die Glia in Hypoglossuskern der Ratte nach Axotomie. Z Zellforsch 141:427–442

Deck JHN, Eng LF, Bigbee J, Woodcock SM (1978) The role of glial fibrillary acidic protein in diagnosis and control of central nervous system tumors. Acta Neuropath (Berl) 42:183–190

Del Cerro M, Swarz J (1976) Prenatal development of Bergmann-glial fibres in rodent cerebellum. J Neurocytol 5:669–676

Diamond MC, Krech D, Rosenzweig MR (1964) The effect of enriched environment on the histology of the rat cerebral cortex. J Comp Neurol 123:111–119

Elliott HC (1969) Textbook of neuroanatomy. JB Lippincott, Philadelphia; Blackwell, Oxford

Fleischhauer K (1966) Über die postnatale Entwicklung der subependymalen und marginalen Gliafaserschichten in Gehirn der Katze. Z Zellforsch 75:96–108

Fleischhauer K (1968) Postnatale Entwicklung der Neuroglia. Acta Neuropathol [Suppl] 4:20–32

Fleischhauer K (1970) Über die postnatale Entwicklung des Stratum subcallosum im Vorderhorn des Seitenventrikels der Katze. Z Anat Entwicklungsgesch 132:1–17

Fleischhauer K (1972) Ependymal and subependymal layer. In: Bourne GH (ed) The structure and function of nervous tissue, vol VI. Academic, New York

Friede RL (1965) Enzyme histochemistry of neuroglia. In: De Robertis EDP, Carrea R (eds) Biology of neuroglia, progress in brain research, vol 15. Elsevier, New York, pp 35–47

Fujita S (1965a) The matrix cell and histogenesis of the nervous system. Laval Med 36:125–130

Fujita S (1965b) An autoradiographic study on the origin and fate of the subpial glioblasts in the embryonic chick spinal cord. J Comp Neurol 124:51–60

Fujita S (1967) Quantitative analysis of cell proliferation and differentiation in the cortex of the postnatal mouse cerebellum. J Cell Biol 32:277–288

Fujita S, Shimada, M, Nakamura T (1966) H^3-Thymidine autoradiographic studies on the cell proliferation and differentiation in the external and internal granular layers of the mouse cerebellum. J Comp Neurol 128:191–208

Fülöp Z, Lakos I, Bascó E, Hajós F (1979) Identification of early glial elements as the precursors of Bergmann-glia: a Golgi analysis of the developing rat cerebellar cortex. Acta Morphol Acad Sci Hung 276:273–280

Garber BB, Huttenlocher PR, Larramendi LHM (1980) Self assembly of cortical plate cells in vitro within mouse cerebral aggregates. Golgi and electron microscopic analysis. Brain Res 201:255–278

Garthwaite J, Woodhams PL, Collins MJ, Balázs R (1980) A morphological study of incubated slices of rat cerebellum in relation of postnatal age. Dev Neurosci 3:90–99

Gaze R, Watson W (1968) Cell division and migration in the brain after optic nerve lesions. In: Wolstenholme GEW, O'Connor M (eds) Growth of the nervous system. Brown, Boston, pp 53–67

Gerschenfeld HM, Wald F, Zadunaisky JA, De Robertis EDP (1959) Function of astroglia in the water-ion metabolism of the cerebral nervous system. Neurology (NY) 9:412–425

Giacobini E, Vernadakis A, Shahar A (eds) (1980) Tissue culture in neurobiology. Raven, New York

Gilmore SA (1971) Neuroglial population in the spinal white matter of neonatal and early postnatal rats. An autoradiographic study on numbers of neuroglia and changes in their proliferative activity. Anat Rec 171:283–292

Glees P, Meller K (1968) Morphology of neuroglia. In: Bourne GH (ed) The structure and function of nervous tissue, vol 1. Academic, New York, pp 301–324

Golgi C (1885) Sulla fina anatomia degli organi centrali del sistema nervoso. Republished in: Opera Omnia, Hoepli, Milan, 1903, pp 397–536

Greenfield JG, Blackwood W, McMenemy WH, Meyer A, Norman RM (1958) Neuropathology. Arnold, London, pp 34–35

Hajós F (1980) The structure of cleft material in spine synapses of rat cerebral and cerebellar cortices. Cell Tissue Res 206:477–486

Hajós F, Woodhams PL, Bascó E, Csillag A, Balázs R (1981) Proliferation of astroglia in the embryonic mouse forebrain as revealed by simultaneous immunocytochemistry and autoradiography. Acta Morph 29:361–364

Hajós F, Feminger A, Bascó E, Mezey É (1982) Transport of horseradish peroxidase by processes of radial glia from the pial surface into the mouse brain. Cell Tiss Res 224:189–194

Hamberger A, Sellström A (1975) Techniques for separation of neurons and glia and their application to metabolic studies. In: Berl S, Clarke DD, Schneider D (eds) Metabolic compartmentation and neurotransmission. Plenum, New York

Hang H (1972) Die postnatale Entwicklung der Gliadeckschicht der Sehrinde der Katze: eine elektronenmikroskopische Studie über die Ausbildung von Lamellenstapeln. Z Zellforsch 123:544–565

Hatten ME, Liem RKH (1981) Astroglial cells provide a template for the positioning of developing cerebellar neurons in vitro. J Cell Biol 90:622–630

Henn FA, Hamberger A (1971) Glial cell function: uptake of transmitter substances. Proc Natl Acad Sci USA 68:2686–2690

Hinds JW (1968a) Autoradiographic study of histogenesis in the mouse olfactory bulb. I. Time of origin of neurons and neuroglia. J Comp Neurol 134:287–304

Hinds JW (1968b) Autoradiographic study of histogenesis in the mouse olfactory bulb. II. Cell proliferation and migration. J Comp Neurol 134:305–322

Hinds JW, Hinds PL (1972) Reconstruction of dendritic growth cones in neonatal mouse olfactory bulb. J Neurocytol 1:169–187

Hirose G, Bass NH (1973) Maturation of oligodendroglia and myelinogenesis in rat optic nerve: a quantitative histochemical study. J Comp Neurol 152:201–210

His W (1887) Die Entwicklung der ersten Nervenbahnen beim menschlichen Embryo: uebersichtliche Darstellung. Arch Anat Physiol Leipzig Anat Abt 92:368–378

His W (1904) Die Entwicklung der menschlichen Gehirns während der ersten Monate. Hirzel, Leipzig

Hommes OR, Leblond CP (1969) Mitotic division of neuroglia in the normal adult rat. J Comp Neurol 129:269–278

Horstmann E (1954) Die Faserglia des Selachiergehirnes. Z Zellforsch 39:588–617

Hösli E, Hösli L (1976) Autoradiographic studies on the uptake of 3H-noradrenaline and 3H-GABA in cultures of rat cerebellum. Exp Brain Res 26:319–324

Hubel DH, Wiesel TN (1962) Receptive fields, binocular interaction and functional architecture in the cat's visual cortex. J Physiol (Lond) 160:106–154

Hubel DH, Wiesel TN (1977) Functional architecture of macaque monkey visual cortex. Ferrier Lecture. Proc R Soc Lond (Biol) 198:1–59

Hutchison HT, Werrbach K, Vance C, Haber B (1974) Uptake of neurotransmitters by clonal lines of astrocytoma and neuroblastoma in culture. I. Transport of γ-aminobutyric acid. Brain Res 66:265–274

Ikeda A, Yoshii I, Mishima N (1980) An immunohistochemical study of the Müller cells of the chicken retina. Arch Histol Jpn 2:175–183

Imamoto K, Paterson J, Leblond CP (1978) Radioautographic investigation of gliogenesis in the corpus callosum of young rat. I. Sequential changes in oligodendroglia. J Comp Neurol 180:115–128

Iversen LL, Dick F, Kelly JS, Schon F (1975) Uptake and localization of transmitter amino acids in the nervous system. In: Berl S, Clark DD, Schneider D (eds) Metabolic compartmentation and neurotransmission. Plenum, New York, pp 65–90

Ivy G, Killackey H (1978) Transient population of glial cells in developing rat telencephalon revealed by horseradish peroxidase. Brain Res 158:213–218

Jacobson M (1978) Developmental neurobiology. Plenum, New York

Karnovsky MJ (1964) A formaldehyde-glutaraldehyde fixative of high osmolality in electron microscopy. J Cell Biol 27:137A

Katchalsky AK, Rowland V, Blumenthal R (1974) Dynamic patterns of brain cell assemblies. Neurosci Res Prog Bull 12:1–187

Kendall JW, Grimm Y, Shimshak G (1969) Relation of cerebrospinal fluid circulation to the ACTH-suppressing effects of corticosteroid implants in the rat brain. Endocrinology 85:200–208

Kendall JW, Jacobs JJ, Kramer RM (1972) Studies on the transport of hormones from the cerebrospinal fluid to hypothalamus and pituitary. In: Knigge KM, Scott DE, Weindl A (eds) Brain-endocrine interaction. Median eminence: structure and function. International Symposium Munich 1971. Karger, Basel, pp 342–349

Kendall JW, McGilvra R, Lamorena TL (1973) ACTH in cerebrospinal fluid and brain. In: 55th Annual meeting of the Endocrine Society, Chicago, (Abstr) p 79

King JS (1968) A light and electron microscopic study of perineuronal glial cells and processes in the rabbit neocortex. Anat Rec 161:11–124

Klatzo I, Piraux A, Laskowski EJ (1958) The relationship between edema, blood-brain barrier and tissue elements in a local brain injury. J Neuropath Exp Neurol 17:548–564

Kölliker A (1896) Handbuch der Gewebelehre, vol 2, 6th edn. Engelmann, Leipzig

Korr H, Schultze B, Maurer W (1975) Autoradiographic investigations of glial proliferation in the brain of adult mice. I. The DNA synthesis phase of neuroglia and endothelial cells. J Comp Neurol 150:169–176

Kreutzberg G (1966) Autoradiographische Untersuchung über die Beteilung von Gliazellen an der axonalen Reaktion im Facialiskern der Ratte. Acta Neuropathol 7:149–161

Kristensson K, Olsson Y, Sjøstrand J (1971) Axonal uptake and retrograde transport of exogenous proteins in the hypoglosal nerve. Brain Res 32:399–406

Kuffler SW (1967) Neuroglial cells: physiological properties and a potassium mediated effect of neuronal activity on the glial membrane potential. Proc R Soc Lond (Biol) 168:1–21

Kuffler SW, Nicholls JG (1966) The physiology of neuroglial cells. Ergebn Physiol 57:1–90

Kuhlenkampf H (1952) Das Verhalten der Neuroglia in den Vorderhörnen des Rückenmarkes der weissen Maus unter dem Reiz physiologischer Tätigkeit. Z Anat Entwicklungsgesch 116:304–312

Lagenaur C, Sommer I, Schachner M (1980) Subclass of astroglia in mouse cerebellum recognised by monoclonal antibody. Dev Biol 79:377–378

La Vail J, La Vail MM (1972) Retrograde axonal transport in the central nervous system. Science 176:1416–1417

Lenhossék MV (1891) Zur Kenntnis der ersten Entstehung der Nervenzellen und Nervenfasern beim Vogelembryo. Verh X Int Med Cong Berl Abth 2:115–124

Léránth Cs, Schiebler TH (1974) Über die Aufnahme von Peroxidase aus dem 3. Ventrikel der Ratte. Electronenmikroskopische Untersuchungen. Brain Res 67:1–11

Levitt P, Rakic P (1980) Immunoperoxidase localization of glial fibrillary acidic protein in radial glia cells and astrocytes of developing rhesus monkey brain. J Comp Neurol 193:815–840

Levitt P, Cooper ML, Rakic P (1981) Coexistence of neuronal and glial precursor cells in the cerebral ventricular zone of the foetal monkey: an ultrastructural immunoperoxidase analysis. J Neurosci 1:27–39

Lewis PD (1968a) A quantitative study of cell proliferation in the subependymal layer of the adult rat brain. Exp Neurol 20:203–207

Lewis PD (1968b) Mitotic activity in the primate subependymal layer and the genesis of gliomas. Nature 217:974–975

Lewis PD, Fülöp Z, Hajós F, Balázs R, Woodhams PL (1977) Neuroglia in the internal granular layer in the developing rat cerebellum. Neuropathol Appl Neurobiol 3:183–190

Lewis PD, Lai M (1974) Cell generation in the subependymal layer of the rat brain during the early postnatal period. Brain Res 75:520–525

Ling EA, Leblond CP (1973) Investigation of glial cells in semithin sections. II. Variation with age in the numbers of the various glial cell types in rat cortex and corpus callosum. J Comp Neurol 149:73–81

Löfgren F (1959) New aspects of the hypothalamic control of the adenohypophysis. Acta Morph Neerl Scand 2:220–229

Lorente de Nó R (1933) Studies on the structure of the cerebral cortex. J Psychol Neurol (Lpz) 48:381–438

Ludwin SK, Kosek JC, Eng LF (1976) The topographical distribution of S-100 and GFA proteins in the adult rat brain: an immunocytochemical study using horseradish peroxidase-labelled antibodies. J Comp Neurol 165:97–208

Lund R, Mustari M (1977) Development of the geniculocortical pathway in the rat. J Comp Neurol 173:289–306

Lynn JA, Panopie IT, Martin JH, Shaw ML, Race GJ (1968) Ultrastructural evidence for astroglial histogenesis of the monstrocellular astrocytoma (so-called monstrocellular sarcoma of brain). Cancer 22:356

Magini G (1888a) Sur la neuroglie et les cellules nerveuses cerebrales chez les foetus. Arch Ital Biol 9:59–60

Magini G (1888b) Nouvelles recherches histologiques le cerveau du foetus. Arch Ital Biol 10:384–387

Mares V, Brückner G (1978) Postnatal formation of non-neuronal cells in the rat occipital cerebrum: an autoradiographic study of the time and space pattern of cell division. J Comp Neurol 177:519–528

Mares V, Lodin Z (1974) An autoradiographic study of DNA synthesis in adolescent and adult mouse forebrain. Brain Res 76:557–561

Maurer W, Schultze B, Schemmer AC, Haack V (1972) Autoradiographic studies on the mode of growth in jejunal crypt cells of the mouse. J Microsc 96:181–189

Meller K, Breipohl W, Glees P (1968) The cytology of the developing molecular layer of mouse motor cortex. An electron microscopic and Golgi impregnation study. Z Zellforsch 86:171–183

Mesulam MN (1976) The blue reaction product in horseradish peroxidase neurohistochemistry: incubation parameters and visibility. J Histochem Cytochem 24:1273–1280

Miale IL, Sidman RL (1961) An autoradiographic analysis of histogenesis in the mouse cerebellum. Exp Neurol 4:277–296

Millhouse OE (1971) A Golgi study of third ventricle tanycytes in the adult rodent brain. Z Zellforsch 121:1–13

Millhouse OE (1972) Light and electron microscopic studies of the ventricular wall. Z Zellforsch 127:149–174

Mitro A (1974) Effect of hydrocortisone injection into the brain ventricles on the adrenocortical reaction to stress in the rat. In: Ependyma and neurohormonal regulation. Int Symp Smolenice 1972, Veda Publishing House of the Slovack Acad Sci, Bratislava, pp 121–134

Mori S, Leblond CP (1969a) Identification of microglia in light and electron microscopy. J Comp Neurol 135:57–80

Mori S, Leblond CP (1969b) Electron microscopic features and proliferation of astrocytes in the corpus callosum of the rat. J Comp Neurol 137:197–205

Mori S, Leblond CP (1970) Electron microscopic identification of three classes of oligodendrocytes and a preliminary study of their proliferative activity in the corpus callosum of young rats. J Comp Neurol 139:1–30

Moskovkin GN, Fülöp Z, Hajós F (1978) Origin and proliferation of astroglia in the immature rat cerebellar cortex. A double label autoradiographic study. Acta Morphol Acad Sci Hung 26:101–106

Mountcastle VB (1957) Modality and topographic properties of single neurons of cat's somatic sensory cortex. J Neurophysiol 20:408–434

Mountcastle VB (1978) An organizing principle for cerebral function: the unit module and the distributed system. In: Edelman GM, Mountcastle VB (eds) The midful brain: cortical organization and the group-selective theory of higher brain function. MIT, Cambridge, pp 7–50

Müller H (1851) quoted by Polyak SL (1956) The vertebrate visual system. University of Chicago Press, Chicago, p 256

Murray M (1968) Effects of dehydration on the rate of proliferation of hypothalamic neuroglia cells. Exp Neurol 20:460–468

Nakai Y, Naito N (1975) Uptake and bidirectional transport of peroxidase injected into the blood and cerebrospinal fluid by ependymal cells of the median eminence. In: Knigge KM, Scott DE, Kobayashi H, Ishii S (eds) Brain-endocrine interaction II. The ventricular system. 2nd Int Symp Shizuoka 1974. Karger, Basel, pp 94–108

Nowakowsky RS, Rakic P (1979) The mode of migration of neurons in the hippocampus: a Golgi and electron microscopic analysis in foetal rhesus monkey. J Neurocytol 8:697–718

Obersteiner H (1883) Der feinere Bau der Kleinhirnrinde bei Menschen und Tieren. Biol Zentralbl 3:145–155

Oksche A (1968) Die pränatale und vergleichende Entwicklungsgeschichte der Neuroglia. Acta Neuropathol [Suppl] 4:4–19

Orkand RK (1977) Glial cells. In: Brookhard JM, Mountcastle VB, Kandel ER, Geiger SR (eds) Handbook of physiology. Section 1: the nervous system, vol I. Cellular biology of neurons, part 2. American Physiology Society, Bethesda, Md., pp 855–875

Palay S, Chan-Palay V (1974) Rapid Golgi method–multiple impregnation process. In: Cerebellar cortex. Cytology and organization. Springer, Berlin Heidelberg New York, pp 333

Palkovits M (1978) Regional distribution of neurohormones in the central nervous system. In: Vincent JD, Kordon C (eds) Cell biology of hypothalamic neurosecretion. Edition du CNRS, Paris, pp 339–356

Palkovits M, Magyar P, Szentágothai J (1971) Quantitative histological analysis of the cerebellar cortex in the cat. I. Number and arrangement in space of the Purkinje cells. Brain Res 32:1–13

Paterson JA, Privat A, Ling EA, Leblond CP (1973) Investigation of glial cells in semithin sections. III. Transformation of subependymal cells into glial, as shown by autoradiography after H^3-thymidine injection into the lateral ventricle of the brain of young rats. J Comp Neurol 149:83–102

Penfield W (1931) The classification of gliomas and neuroglia cell types. Arch Neurol Psychiat 26:745

Penfield W (1932) Neuroglia, normal and pathological. Cytology and cellular pathology of the nervous system. Hoeber, New York, vol 2, pp 421–479

Peters A, Palay SL (1965) An electron microscope study of the distribution and patterns of astroglial processes in the central nervous system. J Anat 99:419

Peters A, Proskauer CC (1969) The ratio between myelin segments and oligodendrocytes in the optic nerve of the adult rat. Anat Rec 163:243 (Abstr)

Peters A, Vaughn JE (1970) Morphology and development of the myelin sheath. In: Davison AN, Peters A (eds) Myelination. Thomas, Springfield, pp 3–79

Peters A, Palay SL, Webster H De F (1970) The fine structure of the nervous system: the cells and their processes. Harper and Row, New York

Podluslo SE, Norton WT (1972) The bulk separation of neuroglia and neuronal perikarya. In: Marks N, Rodnight R (eds) Research methods in neurochemistry, vol 1. Plenum, New York, pp 19–32

Polak M (1965) Morphological and functional characteristics of the central and peripheral neuroglia (Light microscopical observations). In: De Robertism EDP, Carrea R (eds) Biology of neuroglia, progress in brain research, vol 15. Elsevier, Amsterdam, p 14

Polyak SL (1941) The retina. The University of Chicago Press, Chicago

Polyak SL (1956) The vertebrate visual system. The University of Chicago Press, Chicago

Privat A (1975) Postnatal gliogenesis in mammalian brain. Int Rev Cytol 40:281–323

Privat A, Leblond CP (1972) The subependymal layer and neighbouring region in the brain of the young rat. J Comp Neurol 146:277–302

Raedler E, Raedler A (1978) Autoradiography study of early neurogenesis in rat neocortex. Anat Embryol 154:267–284

Raimondi AJ, Evans JP, Mullen S (1962) Studies of cerebral edema. III. Alterations in the white matter: an electron-microscopic study using ferritin as a labelling compound. Acta Neuropathol (Berl) 2:177–197

Rakic P (1971a) Guidance of neurons migrating to the fetal monkey neocortex. Brain Res 33:471–476

Rakic P (1971b) Neuron-glia relationship during granule cell migration in the developing cerebellar cortex. A Golgi and electronmicroscopic study in *Macacus rhesus*. J Comp Neurol 141:283–312

Rakic P (1972) Mode of cell migration to the superficial layers of fetal monkey neocortex. J Comp Neurol 145:61–84

Rakic P (1981) Neuronal – glial interaction during development. Trends Neurosci 7:184–187

Ramón y Cajal S (1890) Sur l'origine et les ramifications des fibres nerveuses de la moelle embryonnaire. Anat Anz 5:85–95, 111–119

Ramón y Cajal S (1904) Sistema nervioso del hombre y los vertebrados. Imprenta y Libreria de N Moya, Madrid

Ramón y Cajal S (1906) Studien über die Hirnrinde des Menschen, vol 5. Barth, Leipzig

Ramón y Cajal S (1909) Histologie du systeme nerveux de l'homme et des vertebres, vol 1. Maloine, Paris (Reprinted in 1952 by Consejo Superior de Investigaciones Cientificas, Inst Ramón y Cajal, Madrid)

Ramón y Cajal S (1911) Histologie du système nerveux de l'homme et des vertébrés, vol 2. Maloine, Paris (Reprinted by Consejo Superior de Investigaciones Cientificas, Inst Ramón y Cajal, 1955, Madrid)

Ramón y Cajal S (1929) Études sur la neurogenèse de quelques vertébrés. VII. Évolution des grains du cervelet. pp 186–188. Izquierdo, Madrid

Retzius G (1893) Studien über Ependym und Neuroglia. Biol Untersuch Stockholm, NS, 5:9–26

Retzius G (1894) Die Neuroglia des Gehirns beim Menschen und bei Säugetiern. Biol Untersuch (Stockh) NS 6:1–24

Reynolds ES (1963) The use of lead citrate at high pH as an electronopaque stain in electron microscopy. J Cell Biology 17:208–215

Reznikov KY (1968) Incorporation of thymidine-H^3 into brain cells of adult mice after brain injury and RNA injection into the brain. Dokl Akad Nauk SSSR 181:467–469 (in Russian)

Reznikov KY, Verbitskaya LB, Kesarev VS, Viktorov IV (1978) Postnatal histogenesis and cell proliferation in the parietal area of the neocortex in mice under normal conditions and after brain injury. Bull Exp Biol Med Akad Nauk SSSR 85:234–237 (in Russian)

Robain O (1970) Gliogenese post-natale chex le lapin. J Neurol Sci 11:445–461

Robain O, Ponsot G (1978) Effects of undernutrition on glial maturation. Brain Res 149:379–398

Rodriguez EM (1972) Comparative and functional morphology of the median eminence. In: Knigge KM, Scott DE, Weindl A (eds) Brain endocrine interaction. Median eminence: structure and function. Int Symp, Munich, 1971. Karger, Basel, pp 319–334

Roessmann V, Velasco ME, Sindley SD, Gambetti P (1980) Glial fibrillary acidic protein (GFAP) in ependymal cells during development. An immunocytochemical study. Brain Res 200:13–21

Rose SPR (1969) Neurons and glia: separation techniques and biochemical interrelationships. In: Lajtha A (ed) Handbook of Neurochemistry, vol 2. Plenum, New York, pp 183–193

Schaper A (1894) Die morphologische und histologische Entwicklung des Kleinhirns der Teleostier. Anat Anz 9:489–501

Schimrigk K (1966) Über die Wandstruktur der Seitenventrikel und des dritten Ventrikels beim Menschen. Z Zellforsch Mikrosk Anat 70:1–20

Schlessinger A, Cowan WM, Gottlieb D (1975) An autoradiographic study of the time of origin and pattern of granule cell migration in the dentate gyrus of the rat. J Comp Neurol 159:149–152

Schmechel DE, Rakic P (1979a) A Golgi study of radial glial cells in developing monkey telencephalon: morphogenesis and transformation into astrocytes. Anat Embryol 156:115–152

Schmechel DE, Rakic P (1979b) Arrested proliferation of radial glial cells during midgestation in rhesus monkey. Nature 277:303–305

Schon P, Kelly JS (1974) The characterisation of ^{3}H-GABA uptake into the satellite glial cells of rat sensory ganglia. Brain Res 66:289–300

Schousböe A, Hertz L, Svenneby G (1977) Uptake and metabolism of GABA in astrocyte cultures from dissociated mouse brain hemispheres. Neurochem Res 2:217–229

Schrier BK, Thompson EJ (1974) On the role of glial cells in the mammalian nervous system. Uptake, excretion and metabolism of putative neurotransmitters by cultured glial tumor cells. J Biol Chem 249:1769–1780

Schubert D (1975) The uptake of GABA by clonal nerve and glia. Brain Res 84:87–98

Schultze B, Gerhard H, Schumpf E, Maurer W (1973) Autoradiographische Untersuchung über die Proliferation der Hepatocyten bei der Regeneration der CCl_4 – Leber der Maus. Virchows Arch [Path Anat] 14:329–343

Scott DE, Krobisch-Dudley G, Knigge KM (1974) The ventricular system in neuroendocrine mechanisms. II. In vivo monoamine transport by ependyma of the median eminence. Cell Tissue Res 154:1–16

Seress L (1978) Divergent responses to thyroid hormone treatment of the different secondary germinal layers in the postnatal rat brain. J Hirnforsch 19:395–403

Seress L (1980) Development and structure of the radial glia in the postnatal rat brain. Anat Embryol 160:213–226

Shoukimas GM, Hinds JW (1978) The development of the cerebral cortex in the embryonic mouse: an electron microscopic serial section analysis. J Comp Neurol 179:795–835

Sidman RL, Rakic P (1973) Neuronal migration, with special reference to developing human brain: a review. Brain Res 62:1–35

Sjöstrand J (1965) Proliferative changes in glial cells during nerve regeneration. Z Zellforsch 68:481–493

Sjöstrand J (1966a) Glial cells in the hypoglossal nucleus of the rabbit during nerve regeneration. Acta Physiol Scand [Suppl] 270, 67:1–17

Sjöstrand J (1966b) Morphological changes in glial cells during nerve regeneration. Acta Physiol Scand [Suppl] 270, 67:19–43

Skoff RP, Price DL, Stocks A (1976a) Electron microscopic autoradiographic study of gliogenesis in the rat optic nerve. I. Cell proliferation. J Comp Neurol 169:291–312

Skoff RP, Price DL, Stocks A (1976b) Electron microscopic autoradiographic studies of gliogenesis in rat optic nerve. II. Time of origin. J Comp Neurol 169:313–334

Smart I (1961) The subependymal layer of the mouse brain and its cell production as shown by radioautography after thymidine-H^3 injection. J Comp Neurol 116:325–347

Smart I, Leblond CP (1961) Evidence for division and transformation of neuroglia cells in the mouse brain as derived from autoradiography after injection of thymidine-H^3. J Comp Neurol 116:349–367

Snodgrass SR, Iversen LL (1974) Amino acid uptake into human brain tumors. Brain Res 76:95–107

Soemmerring ST (1841) Hirn- und Nervenlehre. Umgearbeitet von G Valentin. Voss, Leipzig

Somjen GG (1973) Electrogenesis of sustained potentials. Prog Neurobiol 1:199–237

Somjen GG (1975) Electrophysiology of neuroglia. Ann Rev Physiol 37:163–190

Somjen GG, Trachtenberg M (1979) Neuroglia as generator of extracellular current. In: Speckman EJ, Caspers H (eds) Origin of cerebral field potentials. Thieme Stuttgart

Sotelo C (1978) Maturation of the Nervous System In: Corner MA, Baker RE, Van de Poll NE, Swaab DF, Uylings HBM (eds) Progress in brain research. Elsevier, Amsterdam, 48:149–170

Spacek J (1971) Three-dimensional reconstruction of astroglia and oligodendroglia cells. Z Zellforsch 112:430–442

Stanfield B, Cowan WM (1979) The development of the hippocampus and dentate gyrus in normal and reeler mice. J Comp Neurol 185:423–460

Stensaas L, Gilson BC (1972) Ependymal and subependymal cells of the caudatopallial junction in the lateral ventricle of the neonatal rabbit. Z Zellforsch 132:297–322

Stensaas L, Stensaas S (1968) Astrocytic neuroglial cells, oligodendrocytes and microgliacytes in the spinal cord of the toad. II. Electron microscopy. Z Zellforsch 86:184–213

Sturrock RR (1974a) Histogenesis of the anterior limb of the anterior commissure of the mouse brain. I. A quantitative study of changes in the glial population with age. J Anat 117:17–24

Sturrock RR (1974b) Histogenesis of the anterior limb of the anterior commissure of the mouse brain. II. A quantitative study of pre- and postnatal mitosis. J Anat 117:27–35

Sturrock RR (1974c) Histogenesis of the anterior limb of the anterior commissure of the mouse brain. III. An electron microscopic study of gliogenesis. J Anat 117:37–53

Sturrock RR (1976) Light microscopic identification of immature glial cells in semithin sections of the developing mouse corpus callosum. J Anat 122:521–537

Sturrock RR (1978) Development of the induseum griseum. II. A semithin light microscopic and electron microscopic study. J Anat 125:433–445

Svaetichin G, Negishi K, Fatehchand R, Drujan BD, Selvin de Testa A (1965) Nervous function based on interactions between neuronal and non-neuronal elements. In: de Robertis EDP, Carrea R (eds) Biology of neuroglia. Progress in brain research, vol 15. Elsevier, Amsterdam

Swarz JR (1976) The presence of Bergmann fibres in prenatal mouse cerebellum and its implication in cerebellar histogenesis. Anat Rec 194:543

Swarz JR, Del Cerro M (1977) Lack of evidence for glial cells originating from the external granular layer in mouse cerebellum. J Neurocytol 6:241–250

Swarz JR, Oster-Granite ML (1978) Presence of radial glia in foetal mouse cerebellum. J Neurocytol 7:301–312

Szentágothai J (1975) The "Module-Concept" in cerebral cortex architecture. Brain Res 95:475–496

Szentágothai J (1977) Functionalis anatomia. Medicina, Budapest (in Hungarian)

Taber Pierce E (1966) Histogenesis of the nuclei griseum pontis, corporis pontobulbaris and reticularis tegmenti pontis (Bechterew) in the mouse: an autoradiographic study. J Comp Neurol 126:219–239

Taber Pierce E (1967) Histogenesis of the dorsal and ventral cochlear nuclei in the mouse: an autoradiographic study. J Comp Neurol 131:27–54

Taber Pierce E (1973) Time of origin of neurons in the brain stem of the mouse. Prog Brain Res 40:53–65

Torvik AE, Skjörten F (1971) Electron microscopic observations on nerve cell regeneration and degeneration after axon lesions. II. Changes in the glial cells. Acta Neuropath 17:265–282

Van der Loos H (1963) Fine structure of synapses in the cerebral cortex. Z Zellforsch 60:815–825

Varon SS, Somjen GG (1979) Neuron-glia interactions. Neurosci Res Prog Bull 17:1–239

Vaughn JE (1969) An electron microscopic analysis of gliogenesis in rat optic nerves. Z Zellforsch 94:293–324

Vaughn JE, Peters A (1967) Electron microscopic studies of early postnatal development of fibrous astrocytes. Am J Anat 121:131–151

Vaughn JE, Peters A (1968) A third neuroglial cell type. An electron microscopic study. J Comp Neurol 133:269–288

Velasco ME, Dahl D, Roessmann U, Gambetti P (1980) Immunocytochemical localization of glial fibrillary acidic protein in human glial neoplasms. Cancer 45:484–494

Virchow R (1860) Cellular pathology as based upon physiological and pathological histology. Twenty lectures delivered in the Pathological Institute of Berlin during the months of February, March and April 1858. Churchill, London

Von Bonin G, Mehler WR (1971) On columnar arrangement of nerve cells in cerebral cortex. Brain Res 27:1–9

Wagner HJ, Pilgrim C (1974) Extracellular and transcellular transport of horseradish peroxidase (HRP) through the hypothalamic tanycyte ependyma. Cell Tissue Res 152:477–491

Watson WE (1974) Physiology of neuroglia. Physiol Rev 54:245–271

Weiss DG (1982) Axoplasmic transport. Springer, Berlin Heidelberg New York

Wenger T, Vigh B, Aros B (1966) Mitotic activity in the periventricular substance of the diencephalon of adult rats. Acta Biol Acad Sci Hung 17:175–183

Wise SP, Jones EG (1976) The organization and postnatal development of the commissural projection of the rat somatic sensory cortex. J Comp Neurol 168:313–344

Wise SP, Jones EG (1978) Developmental studies of thalamocortical and commissural connections in the rat somatic sensory cortex. J Comp Neurol 178:187–208

Wolff JR, Rickmann M (1977) Cytological characteristics of early stages of glial differentiation in the neocortex. Folia Morphol (Praha) 25:235–237

Wollemann M (1974) Biochemistry of brain tumours. Publishing House of the Hungarian Academy Science

Woodhams PL, Cohen J, Mallet J, Balázs R (1980) A preparation enriched in Purkinje cells identified by morphological and immunocytochemical methods. Brain Res 199:435–442

Woodhams PL, Bascó E, Hajós F, Csillag A, Balázs R (1981) Radial glia in the developing mouse cerebral cortex and hippocampus. Anat Embryol 163:331–343

Woodhams PL, Hajós F, Bascó E (1982) Lack of correlation between development of radial glia and cortical plate formation in the mouse cerebral cortex. Neuropath Appl Neurobiol 8:241–250

Zimmerman HM (1955) The nature of gliomas as revealed by animal experimentation. Am J Pathol 31:p 1

Zimmerman HM (1957) The natural history of intracranial neoplasms with special reference to gliomas. Am J Surgery 93:p 913

Zimmerman HM (1969) Brain tumours, their incidence and classification in man and their experimental production. Ann NY Acad Sci 159:p 337

12 Subject Index

Synergetics of the Brain

Proceedings of the International Symposium on Synergetics at Schloß Elmau, Bavaria, May 2–7, 1983

Editors: **E. Basar, H. Flohr, H. Haken, A. J. Mandell**
1983. 192 figures. VIII, 377 pages. (Springer Series in Synergetics, Volume 23). ISBN 3-540-12960-X

The broadly applicable, relaxed yet powerful conceptual tools of synergetics led quite naturally to the unique features of this conference proceedings – a broad spectrum of substantive areas reaching from protein physics to cognitive psychology and their use in the development of mathematical languages, ranging from differential topology through nonlinear dynamics to soliton theory. The intimate mixing of real data and abstract theory was such that an integrative, interdisciplinary language of brain function emerged as a genuine possibility. Problems from every level of the nervous system, from the membrane to psychopathology, reformulated in this context with the imaginativeness and rigor usually reserved for mathematical physics, assumed new and fresh appearances. The meeting generated the implicit promise that many of the concerns of synergetics – including the statistical mechanics on non-equilibrium behavior, emergent patterns, bifurcation, broken symmetries, stability theory, universality, and descriptions of scaling and critical phase transitions – would find appropriate roles in the brain sciences. It is probably safe to say that this was the boldest attempt to date at substantive synthesis and abstract mathematical conceptualization in neuropsychobiology and its outcome is very promising.

Neural Coding of Motor Performance

Editors: **J. Massion, J. Paillard, W. Schultz, M. Wiesendanger**
1983. 88 figures, 7 tables. XI, 348 pages. (Experimental Brain Research, Supplement 7). ISBN 3-540-12140-4

The book deals with the analysis of neural processes preceding or accompanying movement sequences. Recent investigations in animals trained to perform motor tasks in response to sensory cue signals have revealed that many different brain structures are active prior to or during execution of a movement. The timing of neural events with respect to the behavioral act is evaluated by statistical analysis of changes in activity of many single neurons of a given structure. Of particular importance is to find out the behavioral parameter which correlates best with the neural event. Such an analysis provides clues to the meaning of the neural processes for distinct aspects of the behavioral act, such as the trajectory or the speed of a movement.

Springer-Verlag
Berlin
Heidelberg
New York
Tokyo

Advances in Anatomy, Embryology and Cell Biology

Editors: F. Beck, W. Hild,
J. van Limborgh, R. Ortmann,
J. E. Pauly, T. H. Schiebler

Springer-Verlag
Berlin
Heidelberg
New York
Tokyo